11/01

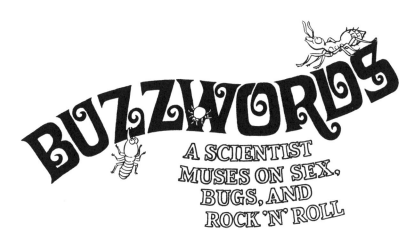

BUZZWORDS

A SCIENTIST MUSES ON SEX, BUGS, AND ROCK 'N' ROLL

May R. Berenbaum

JOSEPH HENRY PRESS
WASHINGTON, D.C.

Joseph Henry Press • 2101 Constitution Avenue, N.W. • Washington, D.C. 20418

The Joseph Henry Press, an imprint of the National Academy Press, was created with the goal of making books on science, technology, and health more widely available to professionals and the public. Joseph Henry was one of the founders of the National Academy of Sciences and a leader of early American science.

Library of Congress Cataloging-in-Publication Data

Berenbaum, M. (May)
    Buzzwords : a scientist muses on sex, bugs, and rock 'n' roll / May R. Berenbaum.
      p. cm.
    Includes bibliographical references (p.).
    ISBN 0-309-07081-3 (hardback : alk. paper) — ISBN 0-309-06835-5 (paperback : alk. paper)
      1. Insects. 2. Insects—Humor. I. Title
    QL463.B47 2000   00-057558

Cover design and interior illustrations by Barbara Spurll

Buzzwords: A Scientist Muses on Sex, Bugs, and Rock 'n' Roll is available from the Joseph Henry Press, an imprint of the National Academy Press, 2101 Constitution Avenue, N.W., Box 285, Washington, DC 20418 (1-800-624-6242 or 202-334-3313 in the Washington metropolitan area; http://www.jhpress.org).

With the exception of Holding the Bag, Kids Pour Coffee on Fat Girl Scouts, An O-pun and Shut Case, Hand-Me-Down Genes, and Subpoenas Envy, all the essays in this volume were previously published in issues of *American Entomologist*. For more information, readers should contact the Entomological Society of America, 9301 Annapolis Road, Lanham, Maryland 20706-3115; www.entsoc.org

Printed in the United States of America

This book is dedicated with love to the memory of Adrienne Berenbaum (May 19, 1918 – December 29, 1999). She was a smart, kind, funny lady who loved science, books, and, above all else, her family. We all miss her very much.

ALSO BY MAY BERENBAUM

Ninety-Nine Gnats, Nits, and Nibblers

Ninety-Nine More Maggots, Mites, and Munchers

Bugs in the System

Contents

BUZZWORDS

HOW ENTOMOLOGISTS SEE THEMSELVES

HOW AN ENTOMOLOGIST SEES SCIENCE

Contents

Preface

Years ago, one of my colleagues taped to his door a cartoon, I think from the *New Yorker*, that made an indelible impression on me (well, perhaps not totally indelible—I either never noticed or didn't bother to remember the name of the cartoonist). The cartoon showed an audience of well-dressed people, obviously at the theater or some similar entertainment, laughing uproariously—all except for a single couple, in the middle of the center row. The man sat, grim-faced and staring, while his wife, with an annoyed expression, admonished him—"For heaven's sake, Stanley, can't you forget for one minute that you're a serious scientist?"

Or words to that effect. The cartoon made an impression because it captures so well the image of the scientist in the public mind. As far as most people are concerned, scientists are the people to whom the rest of the world turns to solve the most intractable problems—disease, hunger, pollution, careening asteroids about to collide with the planet, and the like. And I think it's true that most people who pursue careers in science are motivated by a desire to provide solutions to intractable problems. But the pursuit of these solutions is anything but grim—it's by turns maddeningly frustrating, excruciatingly dull, unspeakably terrifying, and, at wonderful times, utterly exhilarating. I expect it's a lot like many other professions, particularly those with a problem-solving component. While the overall objectives or ultimate goals may be deadly serious, the everyday details generally aren't. Sometimes the everyday details are downright funny (if, on occasion, only in retrospect).

Most scientists don't enter their chosen profession in the hopes of fortune, and few are motivated by a desire for fame. Fortune favors very few and those who really earn vast sums are in a tiny minority. As for fame, even the most famous scientists are generally known only to a handful of people. The exceptional ones, the household names, are few and very far between. Ask anyone to name the first famous scientist who comes to mind and you'll probably hear "Albert Einstein," despite the fact that he's been dead for years. If pressed, most people couldn't even tell you precisely what he was famous for, other than unruly hair and a German accent. Who has to pause to dredge up the name of a famous actor, sports figure, or business magnate who is still breathing?

Why, then, do people become scientists? I can't speak for everyone, but I know why I did. I am a scientist because there is no other activity I can engage in that I find more satisfying. I write these essays in part to share with fellow scientists the joys and frustrations of the business and, as well, in part to show people who aren't scientists just how enjoyable the whole process can be. If these essays don't shatter stereotypes, I hope that at the very least they cause them to crack a little bit.

As enthusiastic as I am about this project, and the overall goal, I have to confess that I didn't come up with the idea in the first place. In 1991, I was invited by Dr. Lowell "Skip" Nault, then president of the Entomological Society of America, to write a "humor column" for the journal *American Entomologist*, a publication of the society that is distributed to all 7,000+ members as a consequence of paying dues. I'm not really sure how Dr. Nault decided on me for this task, but I do know, when he wanted a plenary speaker for the 1991 annual meeting who would inject some humor into the gathering, that he asked me only after Gary Larson, of cartoon fame, turned him down. I was a little hesitant at first, among other things because I wasn't exactly certain that what I considered humorous would strike other entomologists the same way. After nine years, though, I feel more confident that there are some things that are more or less universally funny.

This book, then, consists mainly of columns I wrote for the *American Entomologist* between 1991 and 1999, along with a few additional essays written expressly for this project on topics that are less specifically of relevance to entomologists. The essays written for the *American Entomologist* have been adapted for this book by expeditious purging of unnecessary jargon and entomological inside jokes. All of the essays fall into four major categories, corresponding to the sections in the book. The first section, "How entomologists see insects," consists of essays about the insects themselves; it was, in most cases, the amazing details of the lives of these incredible creatures that got most of us entomologists interested in the field in the first place. The second section, "How the world sees insects," deals with the prejudices of the public at large toward insects—on occasion a source of great amusement, although more frequently a cause of great consternation, to entomologists. The amusement arises at least in part because, despite their avowed dislike for insects, people have found some remarkable ways to incorporate insects and their images into their daily routine. The concern arises from the tendency of people to believe the worst about insects, no matter how outlandish or dangerous those beliefs may be. The third section, "How entomologists see themselves," contains essays on how difficult it can be to explain entomology as a career choice to the world at large—a reflection of how difficult it is to explain the scientific enterprise in general to people whose last experience with scientific research was a mandatory one-semester general education class in college (which they hated). Finally, the fourth section, "How an entomologist sees science," has to do with the business of science irrespective of discipline—these essays address the commonalities of the conduct of science despite attempts within the scientific community to differentiate and classify scientists according to discipline, age, gender, research organisms, experimental approaches or whatever.

So, of entomologists, the general public, and the scientific community, I don't really know who will end up reading this book. Whoever you are, please enjoy these essays—they're written to be enjoyed. But before you proceed, here's a word of warning. In these essays, you'll encounter scientific names. For reasons I'm not entirely clear on, these seem to alarm people, even some biologists, unnecessarily. These names, which are written in Latin and consist of two parts, the genus followed by the species, are used not to impress people with dazzling displays of arcane knowledge; I don't know that I've ever won anyone's heart or stopped a fight or brought the world one step closer to peace and tranquility by reeling off a scientific name at a critical juncture. They're used simply because they're really very useful. For one thing, they're universal; no matter what it may be called in French, German, Italian, Tagalog, or Swahili, each species has only one scientific, or

Latin, name everywhere in the world. Entomologists find them particularly useful because there are many species that just aren't common enough to have names in French, German, Italian, Tagalog, or Swahili. If they bother you, you can gloss over them the way that I used to gloss over all of the names I couldn't pronounce in the Russian novels I read in college.

Coincidentally, one novelist of Russian extraction, Vladimir Nabokov, of "Lolita" fame, happened to be very good at scientific names; he personally bestowed scientific names on several species of gossamer-winged butterflies before he began his literary career in earnest. Nabokov's range of talents reinforces the main idea here—that scientists are more than guys in white lab coats holding Erlenmeyer flasks. Not to give anything away, but you won't find another reference to Erlenmeyer flasks in the rest of the book; I hope you find what is mentioned a lot more interesting.

Acknowledgments

There are many, many people to whom I owe thanks and without whom this book would not exist. First of all, I should thank the 7,000 or so members of the Entomological Society of America (forgive me if I don't list all of your names); many regularly read the Buzzwords columns I write for the *American Entomologist* and have taken the time to let me know how much they enjoyed them, thereby encouraging me to go ahead with this project. I also have to thank the many people who made a point of pointing out the things they'd come across that struck them as funny—many of the ideas for these essays came from letters, cards, e-mail messages, phone calls, or even old-fashioned face-to-face conversations with colleagues. I've tried to give credit where credit is due, but if I've inadvertently left out your name I apologize. And I thank the many authors of articles who shared reprints and photocopies with me that I couldn't otherwise locate. (I'm still looking for the full text of that article by Yuswadi in the 1950 *Siriraj Hospital Gazette*, though.)

Thanks are due, too, to the many entomologists who enthusiastically contacted me whenever I made an error; I really do appreciate these efforts and have tried mightily to expunge

BUZZWORDS

typographical errors, outdated information, and other carelessness from this book. I must assert emphatically, however, that if errors remain, they are certainly not the fault of the Entomological Society of America—they are all my own and I'm not proud of them. I hope you'll contact me without delay if you find some in this book. Although it's embarrassing to get caught in an error, I believe it's a far greater mistake to knowingly leave an error uncorrected—if you find one, I'll do my best to fix it. I also owe the Entomological Society of America thanks for granting me permission to reprint many of the essays with a minimum of fuss.

I also have to thank Stephen Mautner and the staff at Joseph Henry Press for taking a chance on *Buzzwords*. Steve's enthusiasm for the project never flagged, even when mine slipped a little, and he also displayed remarkable tolerance for my poor grasp of the concept of "deadline." Ann Merchant spared no effort in promoting the book and even managed to convince me that being an entomologist was pretty cool. I hope their faith in this book proves to be justified.

Closer to home, I have students, staff, and faculty here in the Department of Entomology at the University of Illinois to thank for generously sharing their expertise with me, and for not even blinking when I sought clarification on subjects that can only be considered a bit odd (colleague Susan Fahrbach, for example, provided me with references to help me understand the phenomenon of reflexive penile erections during paradoxical sleep without so much as a by-your-leave). Several undergraduate assistants were utterly indispensable to me, providing cheerful service searching for and retrieving bits of information from places on campus that entomologists don't usually go. Jodie Ellis was particularly noteworthy in this regard and never hesitated to search

Acknowledgments

the four corners of the enormous UIUC library to hunt down sources, even, for example, when I sent her to locate the decidedly unentomological *Spalding Official Baseball Guide of 1900*. I thank Pete Ficarello too. I must also thank Dr. Arthur Zangerl, with whom I have worked closely for close to 20 years, for too many things to enumerate; in particular, his willingness to apply his statistical expertise to unusual forms of data has been greatly appreciated.

It's perhaps understandable that my colleagues in the department were so willing to help—after all, they've known me for some time and knew about my penchant for writing on nontraditional themes. But I owe a great debt to my colleagues here at UIUC in other departments, who proved to be outstanding sources of obscure information (e.g., how bees buzz in Hindi) and who never failed to offer that information freely and enthusiastically. The library personnel deserve a special sincere and heartfelt thanks for their extraordinary effort on my behalf—piecing together incomplete references, answering questions about where to find things, and demonstrating extraordinary tolerance for my tendency to overlook due dates.

Even closer to home are the people at home—this book wouldn't exist without the support and encouragement of my family. My parents, Morris and Adrienne Berenbaum, have always taken great pride in my accomplishments, such as they have been. They were both trained as scientists and they made sure by their words and actions that science was as much a part of my daily life as breakfast cereal. Thanks to their efforts, it never occurred to me that science was boring and in many ways I owe my career to their example; they will always inspire me. My sister Diane and my brother Alan have always been willing to share with me their memories of our childhood together and as well to share with me their diverse and interesting experiences as grownups. They keep me ever mindful of the world beyond entomology and help keep me in touch with forms of culture other than microbial. And, of course, I thank Richard Leskosky for sharing his life with me. His influence permeates this book; he read these essays long before I would have dared show them to anyone else and helped to make them presentable to others. That he always knows the right words to say, whether they're the words I need to fix an awkward title or finish off an endless essay, or to shake off despair and see the humor in things, is not so much a reflection of his training in linguistics and English literature as it is a reflection of his incredible intellect and his generous, good nature.

And I thank my daughter Hannah, for continuing to bring happiness into my life in exponential fashion. Over the years, she has developed not only a fairly sophisticated knowledge of insects (for a nine-year-old at least), but also a remarkable sense of humor. She cracked her first joke at the amazingly precocious age of two, and the jokes, wry remarks, and witty insights keep getting better. And turnabout is fair play. Making her smile isn't always easy—she has never been an easy audience, particularly for ponderous parental attempts at humor—but hearing her laughter is a joy beyond comparison.

How
entomologists
see
insects

On elderly ants

Most of the owners of secondhand bookstores in town recognize me on sight and inevitably greet me with a smile whenever I walk in. The smile is more than just good retail business practice—they know that odds are excellent that when I leave I will take with me a lot of merchandise and leave behind a lot of money. I don't collect things as a rule—not coins, not matchbook covers, and not even insects to speak of—but I do seem to have a compulsive need to own out-of-date, cracked, yellowing books about insects. No matter how long that copy of *Entomological Papers from the Yearbook of Agriculture 1903-1911* has sat moldering on the shelf—the booksellers know once I walk in they'll never have to dust it again. No matter how ridiculously overpriced that 1910 copy of *How to Keep Bees for Profit* is—the checkbook will open and the ink will flow. Sometimes in their zeal, these booksellers will show me books about snakes, worms, snails, and other noisome creatures, but to date I have usually managed to contain my impulses, succumbing only if there's a passing reference within the volume to anything six-legged.

As hobbies go, this one isn't bad, really—it's legal, it's not as expensive as, say, powerboat racing or big game hunting (not to mention a lot safer), and, best of all, it's not fattening. This hobby is how I happened to read Lord Avebury's account of a 14-year-old ant. A visit to Old Main Book Shoppe on Walnut Street in downtown Champaign produced a dusty copy of Lord Avebury's *Ants, Bees, and Wasps*, from 1916, which, of course, I bought without even opening. By the way, Old Main Book Shoppe was at one time actually located on Main; I assume the owner just liked the sound of "Old Main" more than "Old Walnut."   Once I got home, I started thumbing through the book and soon came across a passage discussing the life expectancy of ants, which, it happens, was a subject of some controversy a century ago:

> The life of the queen and workers is much longer than had been supposed. I may just mention here that I kept a queen of *Formica fusca* from December 1874 till August 1888, when she must have been nearly fifteen years old, and of course she may have been more. She attained, therefore, by far the greatest age of any insect on record. I have also some workers which I have had since 1875.

When you think about it, keeping an ant alive for 14 years is quite a remarkable feat. After all, it's not like keeping a dog alive for 14 years. For one thing, it's hard to lose or misplace a dog on your desk, and it's even harder to flatten one accidentally under a coffee cup. And it's not like there's a tremendous support system out there for ant owners—all-night ant veterinary services, for example, or ant toys and treats at the local grocery store. I doubt that too many small animal clinics, name notwithstanding, will see ant patients. And it's a lot harder to ignore a dog if you have forgotten to feed it or take it for a walk. What makes this feat even more impressive is that Lord Avebury was an extremely busy guy who had many things to do in life other than look after a geriatric ant. Before becoming the Right Honorable Lord Avebury, he was

4

Sir John Lubbock, DCC, LLD, MD, FRS, VPCS, FGS, FZS, FSA, and FES. He belonged to no fewer than two dozen scientific societies, spread out among seven countries on three continents. I can't help wondering what friends or relations he might have prevailed upon to look in on his ant while he was making the rounds of scientific meetings.

It's no wonder that claims for insect longevity records are few and far between. They're not nonexistent, though. Not long after Lord Avebury's book came out, Ferris (1919) reported finding a single nymph of *Margarodes vitium* alive in a waxy cyst some 17 years after the specimen was deposited in the Stanford collection of Coccidae. E. Gorton Linsely (1943) reported 12 instances in which *Buprestis aurenta*, a woodboring beetle, emerged from structural wood in walls, floors, doors, and stairway handrails, anywhere from 10 to 26 years after the structures were built, in some cases struggling through linoleum to do so. Most recently, Jerry Powell (1989) reported that more than 180 adult yucca moths emerged from cocoons dating back to 1969, after 16 to 17 years in a quiescent diapause state. These cocoons had accumulated a lot of miles, moving from their childhood home in Nevada, to the Berkeley campus for a year, and then to the University of California Russell Reserve in Contra Costa, California. Those cocoons that had not produced an adult by 1985 were partitioned for a while among an outdoor cage at the Russell Reserve, an outdoor cage at Blodgett Forest in El Dorado County, California, and a mobile laboratory on the Berkeley campus. Eventually, all were reunited back in Berkeley for the final emergence.

It would be difficult to judge which feat was more impressive—keeping track of 180 yucca moth larvae for 17 years, or keeping an adult ant alive for 14 years. On one hand, the diapausing larvae don't need to be fed and an adult ant does; on the other hand, while an adult ant is fairly responsive. it would be an almost overwhelming temptation to cut through the cocoons every five years or so just to see if their occupants were still alive. I am fairly certain I could not have accomplished either feat. I can't even keep track of a Sharpie marker on my desk for more than a week.

It must be said at this juncture that in none of these studies of insect longevity did the investigators tamper with the processes of nature. Studies aimed at prolonging the lifespan of insects don't figure prominently in most entomology programs, the vast majority of which have exactly the opposite goal. But there is one group of scientists who for years have done everything they can to make insects live longer. If you haven't been reading journals such

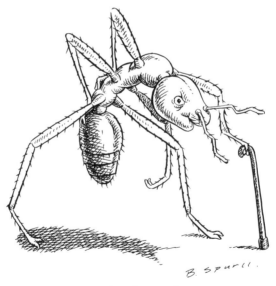

Ant-iquated

as *Age* or *Experimental Gerontology*, then you may not have seen these studies. It turns out that there are many people who test theories of aging with insects. This is not surprising from an experimentalist point of view, when you think about it; practically speaking, it's nice to be able to detect a 50% increase in lifespan when that increase translates to a few days. Comparable studies of long-lived Amazon parrots or Galapagos tortoises could run a century and a half or longer, which far exceeds the average funding cycle of most federal agencies. From this perspective, *Clunio maritimus* could be the ideal subject—the so-called one-hour midge has an adult lifespan of about an hour, give or take 30 minutes. Most people in this area, however, use other flies, mostly *Drosophila melanogaster*, which live a positively Methusaleh-like month or more as adults. Among the myriad substances tested and found to prolong the lifespan of these flies are cortisone, hydrocortisone, aspirin, triamcinolone), meclofenoxate, sodium thiazolidine-4-carboxylate, 2-ethyl-6-methyl-3-hydroxy-pyridine, ethidium bromide, lactic acid, diiodomethane, sodium hypo-phosphite, vitamin E, and innumerable others too obscure to mention.

One has to wonder what Lord Avebury would have thought about using artificial means to prolong insect life. Would he have resorted to any means possible to extend the lifespan of his ant? How long can an ant actually live, if assisted? Evidently, it's still very much an open question. With all of this newfound gerontological knowledge in hand, I just might go ahead and try to answer that question with a study of my own. I'd start right away, too, except that I need to write down a few things first and I can't seem to find my Sharpie marker. . . .

Putting on airs

Despite all human pretenses at being superior to other living things, there are a few body functions around that remind us that we, too, are fundamentally similar to the lower life forms. One such body function is the occasional need to eliminate accumulated waste gases from the digestive tract. In humans, these gases tend to build up in the gut lumen as the result of bacterial fermentation; hydrogen gas is generated as a byproduct of the fermentation of ingested carbohydrates and amino acids, and methane by bacterial fermentation of endogenous material. Although some of these gases can diffuse from the lumen to the bloodstream, they are more frequently expelled at the terminus of the digestive tract by a process known by medical professionals as "flatus." Although it's a legitimate physiological process, the act of expelling excessive intestinal gas for some reason is often regarded as comical. Even the normally staid and stolid *Merck Manual of Diagnosis and Therapy,* a 2,578-page tome documenting in excruciating and often horrific detail every kind of defect human flesh is heir to, lightens up on the subject (p. 793):

> Among those who are flatulent, the quantity and frequency of gas passage can reach astounding proportions. One careful study noted a patient with

daily flatus frequency as high as 141, including 70 passages in one 4-h period. This symptom, which can cause great psychosocial distress, has been unofficially ... described according to its salient characteristics: (1) the 'slider' (crowded elevator type), which is released slowly and noiselessly, sometimes with devastating effect; (2) the open sphincter, or 'pooh' type, which is said to be of higher temperature and more aromatic; and (3) the staccato or drum-beat type, pleasantly passed in private.

Among the handful of people who regard the release of intestinal gas not as a matter for humor in questionable taste but rather as a matter of urgent global concern are scientists who study this universal phenomenon in insects. This is a position that in some respects is well-taken. After all, there are many more insects than there are humans, and mass release of methane, a known greenhouse gas capable of affecting global climate, from so many abdomens (with so many orifices) could have potentially earth-shaking consequences. It's not surprising, then, that much serious study has been devoted to the subject.

The debate about the significance of insect flatulence has been waged for over three-quarters of a century not in obscure special interest entomological journals but rather in the premier scientific journals of our era. It all more or less began in 1923, with L. R. Cleveland's observation that the protozoans living in the guts of termites may actually be performing a useful function from the termite perspective. That this function somehow involved methane was suspected early on but wasn't confirmed for another fifty years. Once it was confirmed, however, quantifying that methane production became a national priority. In 1982, a collaborative effort among four scientists from three continents produced the first estimate of the annual production of methane by termites. In a paper in *Science*, P. R. Zimmerman and colleagues reported measuring carbon monoxide, carbon dioxide, methane, hydrogen, and several short-chain hydrocarbons emitted from the anus of three different termite species and scaling up to the global level from there. They found that individual termites were capable of producing 0.24-0.59 micrograms of methane per day; drawing from literature estimates of the global population densities of termites, they calculated that the approximately 2.4×10^{17} termites in the world could produce 1.5×10^{14} Terragrams methane each year, give or take a few Terragrams, one Terragram being the equivalent of 10^{12} g, a unit of mass not usually dealt with by entomologists. This amount is impressive in its own right, but it's even more impressive given that the annual production of methane from all sources globally was estimated at only 3.5 to 12.1 $\times 10^{14}$ g. In other words, termite flatulence might be responsible for as much as 30% of the earth's atmospheric methane levels—levels that are rising even higher, according to Zimmerman et al., because deforestation and agriculture tend to favor the buildup of termites.

Not long after, in 1983, R. A. Rasmussen and M.A. K. Khalil published a dissenting view in the pages of *Nature*. Based on extrapolations of their own laboratory studies with *Zootermposis angusticollis*, they estimated termite methane production at a mere 5×10^{13} grams per year. These investigators qualified their findings even further by saying there was too much uncertainty not only about how much methane is produced by termites but also about how many termites there are in the world producing methane to come up with any definite figure for global emissions. Zimmerman and Greenberg (1983) were quick to respond, pointing out that, among other things, Rasmussen and Khalil had done their studies with termites confined in sealed and not flow-through containers, which likely affected their findings. Zimmerman and Greenberg actually obtained *Z. angusticollis* from Rasmussen and Khalil and repeated the experiment with their own chambers, obtaining estimates of methane production 3 to 6 times higher than the ones reported by Rasmussen and Khalil. Khalil took the lead in the next response (Khalil and Rasmussen, 1983), reiterating that termite colonies were more like sealed flasks than flow-through chambers and that the original Zimmerman et al. estimate was still probably too large by a factor of three. Another paper in *Science* came out the following year, in which N. M. Collins and T.G. Wood (1984) took issue with the way Zimmerman et al. had interpreted an earlier paper (by Wood and his colleague W.A. Sands, published in 1978) in estimating the total number of termites in the world and also disputed the assumption that deforestation increases termite abundance. Despite the authority that someone with the name of Wood would appear to have on the subject of termites, Zimmerman and colleagues (1984) replied to this criticism as well, acquiring a new co-author in the process. They responded that Collins and Wood had misrepresented their interpretation of the earlier paper, and then proceeded to cite the earlier paper (by Wood and Sands) to support their original estimate of termite densities.

While all of this wrangling was going on in the pages of *Nature* and *Science*, the methane levels of the atmosphere were apparently dropping. This decline did not go unnoticed by Steele et al. (1992), who reported the slowdown, in *Nature* of course. Meanwhile, on the termite front, several laboratories, seemingly oblivious to the waning urgency of the work, were intensively refining the estimates of global methane production by termites. Brauman and colleagues (1992) reported in *Science* that methane production is a function of diet, with rates of methane emission greatest in soil feeders, followed by fungus feeders; wood-feeders bring up the rear, so to speak. Another refinement on the estimates was geographical; Martius et al. (1993) pointed out that the past decade of methane emission measurements were all from North America, Africa, and Australia. According to their findings, termites from North America and Australia couldn't hold a candle to Amazonian termites when it comes to producing methane (although holding a candle to any methane-producing body is probably a bad idea). They estimated that termites of all nationalities were collectively responsible for only 5% of the annual global methane flux.

In 1994, J. H. P. Hackstein and C. K. Stumm published what might be the definitive paper on arthropod methane emission and in doing so raised a terrifying specter. They surveyed more than 100 species of terrestrial arthropods and reported some findings that suggest termites may not be our chief concern in the methane arena. Whatever their other shortcomings, cockroaches are no slackers when it comes to producing methane. All of the major domiciliary species of cockroaches—the German, the Oriental, the American, and the brown-banded—produce in excess of 31 nanomoles per gram fresh weight per HOUR, with *Periplaneta americana*, the American cockroach, pumping out as much as 255 nanomoles per gram fresh weight per hour. While it's true that termites still can outproduce cockroaches by almost twofold on the global scale (50.7 Tg/year globally compared to only 28 Tg/year), with a few exceptions they do most of this methane release in exotic localities such as African savannahs, Australian deserts, and Amazonian rain forests. Cockroach methane production hits just too close to home. It's one thing to have to take the blame for the lingering odor of the digestive upset of kitchen vermin, but it's something else entirely to risk life and limb by cohabiting with cockroaches. Along with methane, cockroaches are capable of producing carbon monoxide, leaving open the possibility that fires of suspicious origin and carbon monoxide poisonings may have to do a lot less with faulty stoves and a lot more with windy cockroaches.

But maybe concerns about arthropod emissions are being blown out of proportion. After all, it's been known for years that at least some species of cockroaches can produce methane; D.L. Cruden and A.J. Markovetz (1984) first quantified the phenomenon a decade earlier and even reported finding a methanogenic bacterium in the gut of a cockroach uncomfortably similar to one originally identified from human excrement (suggesting that humans and cockroaches may be more biologically similar than is pleasant to contemplate). For that matter, there's evidence that cockroaches in particular and insects in general have shared this less than endearing human physiological foible for centuries. *The Florentine Codex*, a "General History of the Things of New Spain," was translated from the Aztec in the sixteenth century by Fray Bernardino de Sahagun; Book 11 (Earthly Things) provides remarkable insights into the level of appreciation for natural history in pre-Columbian Mexico. Part twelve of Book 11 contains an account of the "pincatl. . . . It is blackish, dark, small and flat, with pointed jaws. It is sherd-like; rigid is its sherd. And when anyone molests it, then it breaks wind; it frightens one with its stink, its flatulence. It lives, it dwells, in damp places, in rubbish." We may as well accept it, in ourselves as in others, since it's obviously been going on for a long time without any document-

BUZZWORDS

able disastrous consequences for our species. Tiny gas bubbles are even visible in Dominican amber, clinging to the abdomens of termites, cockroaches, millipedes and other gassy arthropods. The process has been silent for millions of years, but it hasn't proved deadly yet.

Fatal attractions

If you went to public high school in the early seventies, as I did, then you probably are a survivor of an educational experiment called "health class." I took health class in ninth grade, as did every ninth grader in Pennsylvania, because the powers-that-be in the state decreed that all ninth graders had to take health class in order to receive a high school diploma. The curriculum was designed to acquaint us with the hazards of sex, drugs, and inadequate personal hygiene. As I recall, we saw a lot of movies and filmstrips that I think were intended to frighten us. They couldn't have been exceedingly effective because I don't remember much about them. What I remember most clearly about health class was the textbook, and I remember that because the nameless student who had used that particular book the year before it was assigned to me had been thoughtful enough to pencil in all of the obscene terms for the various parts of both male and female reproductive tracts, very few of which I knew before, right next to all of the technical terms. Thus, overall, I have to say I found the class dull, but definitely educational.

Driver education class, on the other hand, was terrifying. We saw movies in that class, too, but these films were so frightening that I ended up literally not getting behind the wheel of a car for eleven years after passing the course and getting my license. For the most part, these films depicted unremittingly horrific scenes of highway carnage. Time has dimmed these memories somewhat, and I do drive on occasion around town, but, thanks to a film called "Signal 30," produced, I think, by the Ohio Traffic Safety Bureau and depicting all manner of gore-filled accidents (including one involving a truckload of cattle), I don't think I'll ever have the courage to drive in the state of Ohio (or eat hamburger, for that matter).

It seems to me that secondary school educators in the state of Pennsylvania missed a golden opportunity to instill morality through terror merely by virtue of their choice of films. I don't know if they still teach health class to ninth graders in Pennsylvania but if they do I recommend that they show a few nature documentaries instead of the movies they showed us. A brief glimpse into the reproductive habits of insects would be enough to put anyone off sex for a long time. Take the courtship ritual of *Calopteran discrepans*, for example. J. M. Sivinski (1981) describes a presumably typical encounter:

> Males mount dorsally, between the females' slightly spread wings. Examination of coupled pairs showed the sickle-like male mandibles bite through the humeral angle (shoulder) of the female's right elytron. Up to 3 males were found upon a female's back. When such masses were picked up they clung together and were separated only with some effort, leaving bleeding wounds in the female's elytra.

In what can be described only as masterful understatement, Sivinski observes that such "love bites . . . illuminate the different reproductive interests of the sexes."

This sort of mangling actually appears to be fairly routine

among insects. In a study of a dozen species of Nearctic gomphid dragonflies, for example, 88 to 100% of the females examined "had 2-6 holes in their heads resulting from the grip of male abdominal appendages" (Dunkle, 1991). The gomphids are apparently far rougher than aeshnid dragonflies, the male of which merely "gouges the dorsal surface of the female's compound eyes." *Hagenius brevistylus*, North America's largest gomphid, earns distinction of a sort by exhibiting "the most severe head damage due to mating attempts so far discovered in any dragonfly." In this species, "the laterodistal spines of the male epiproct gouged the edge of the female's compound eyes, and punctured the exoskeleton in . . . 32% of the females in which the male cerci also punctured the head. A proximodorsal ridge on each side of the male epiproct often . . . cracks the lateral corners of the female occiput. Finally, a distal spine and a mediolateral spine on each male cercus punctures the rear of the female head (postgenae). The pressure of the male grip splits the exoskeleton between the holes made by the cercal spines, resulting in a vertical split in each postgena. Thus a maximally damaged female would have 6 holes of varying sizes punched in her head."

I expect a young, impressionable female high school student who grows up associating words like "gouge," "puncture" "split" and "punch" with the act of copulation might never yield to temptation, even after years of marriage.

Actually, the girls have it relatively easy in the insect world—though they may be disfigured for life, at least they survive these encounters. There are innumerable accounts of sexual encounters among insects that leave males dead. I don't just mean those stories about praying mantids, which may be somewhat exaggerated (see "A prayer before dining" for details). Male *Tribolium* beetles die a particularly horrible kind of death when they're maintained in all-male groups. Male *T. castaneum* beetles kept with females live on average 50 weeks; those in all-male groups die after only 15 weeks. These males die with a hard whitish plug at the tip of their abdomen, the apparent result of solidification of seminal fluids upon contact with air. When food particles adhere to the fluids, what results is a solid mass that interferes with the various and sundry functions of the nether end of the abdomen. Then there's the sad fate of *Julodimorpha bakewell*, a species of buprestid, or flat-headed boring beetle, in Australia. These shiny beetles mistake the shiny surface of a 370-ml beer bottle (called a "stubbie") for a female buprestid and attempt to mate with it, invariably with less than satisfying results. Gwynne and Rentz (1983) conducted a short experiment by placing four bottles on the ground: within thirty minutes, six beetles had arrived to hit on the bottles. The problem with hazards of this behavior is that the beetles don't give up; one male apparently died as the result of attack by ants, "biting at the soft portions of his everted genitalia."

It might be argued that knowledge of these fatal attractions would be of little relevance to people—that small, crawling animals have little to do with human sexual practices. Remarkably enough, that's not always the case. There is an unusual convergence of sex, invertebrates, and humans in a form of fetishist known as the crush-freak.

To define "crush-freak," I refer to the definitive source on the subject, the *American Journal of the Crush-Freak* (1993):

> This is a very unique sexual fantasy, which is part of the foot-fetish. In the "Crush-Freak's" mind he wishes himself tiny—insect like—and wants to be stepped on and squashed by the foot of a woman. There are a number of variations on this fetish fantasy. Some of us want only to be stepped on

barefoot, some only want to be crushed under the pump of the shiny, high-heeled shoe. . . . Others want to create scenarios in which the female imagines that one male is a bug, and gets her boyfriend to stomp him. Many fetishists must see a female step on a tiny living thing, an insect makes a fine surrogate for the "Crush-Freak."

The *American Journal of the Crush-Freaks* is edited by Jeff Vilencia, an aspiring film-maker and self-avowed crush-freak who, in the biographical information appearing in his journal, admits to fantasizing about "being a bug." I learned of Jeff Vilencia and of his unusual entomological interests when he called me up after reading about our departmental Insect Fear Film Festival in an article in *Modern Maturity Magazine* (official publication of the American Association of Retired Persons) at his mother's house. Jeff was kind enough to send me a copy of his award-winning short film, "Smush," approximately eight minutes of actress Erika Elizondo crushing earthworms first with bare feet and then with her mother's black, stiletto-heeled pumps. This film was recognized at the Toronto International Film Festival in 1993, the Helsinki Film Festival of 1994, and, somewhat less surprisingly, at the Sick and Twisted Film Festival of 1995, and was written up in the *New York Post* and the *Washington Post*.

Technically speaking, this sort of sexual encounter really has adverse consequences only for the small invertebrate so I suppose "Smush" really wouldn't be suitable for showing to adolescents to demonstrate to them the hazards of unprotected sex. To be honest, I'm not exactly sure just what the appropriate audience would be for this film. I feel a little bad that I have trouble appreciating the aesthetics of the film because Jeff Vilencia certainly appreciates entomologists. In fact, in his journal he even reviews and rates entomological publications. Of course, he doesn't use the same criteria that, say, I might use in reviewing such a text; his "criteria for inclusion" are that the books must be written by a woman and "that there must be one or more good 'Crush' references."

The issue of the *American Journal of the Crush-Freaks* in my possession contains two such book reviews. Of *Bug Busters*, by Bernice Lifton (1991), Vilencia excerpts six references to insect-crushing, with annotations. Such annotations are often quite succinct—e.g., "p.212, Ch. 13 ANTS, SPIDERS, AND WASPS . . . QUOTE: 'Try to kill the biting spider without squashing it beyond recognition . . . <u>OKAY</u>." This is not to say he's entirely uncritical in his praise of entomological texts, though. In his review he states his disappointment that the author uses the term "squash" instead of "squish"; evidently he finds "squish" a more evocative term. He concludes the review with the note, "we can only hope that Ms. Lifton is a young sexy babe with a size 9 or 10 shoe, and loves to step on bugs!"

Rhonda Wassingham Hart also received accolades from Jeff Vilencia for her book *Bugs, Slugs & Other Thugs*. Vilencia's review ends with what must be the ultimate praise for an entomological text—"I can only say that I would love to be a bug in her garden so she could step on me."

I guess the point of all of this is that what is erotic is largely in the mind, not only for high school kids but also for grown-ups. My experience with Jeff Vilencia has led me to wonder about who reads the books I've written and what motivated them to buy the books in the first place. On the one hand, it's almost gratifying to think that insect pest management can arouse people's interests to such an extreme extent. On the other hand, it has convinced me not to list my shoe size in the biographical sketch of my next book.

Just say "Notodontid?"

The first place I read about the U.S. government's plan to eradicate illicit coca fields by dropping caterpillars from airplanes was not on the front page of our local newspapers—it was farther back, in the editorial section. A spate of editorial cartoons appeared, generally depicting drug czar William Bennett in a number of less-than-flattering ways. Imagine, if you will, the illustration accompanying the caption, "Disguised as a parachuting caterpillar, Wily Coyote Bennett prepares to pounce on his prey, the crafty drug-runner" (Oliphant, Universal Press Syndicate). Or the one where Bennett, crouched in an airplane hold, is dumping moths out the bay door and pointing at crates of benzene-contaminated Perrier, saying "See . . . first we drop the moths on the coca plants . . . and if that doesn't work . . ." (Mike Keefe, the *Denver Post*). Then there's the one that simply reads "Has this gotten stupid enough for you?" (Toles, Universal Press Syndicate).

I was, of course, interested in finding out the real story behind the editorial cartoons. The hometown paper, the *Champaign-Urbana News Gazette*, was of no help at all—I couldn't find any trace of the story. Of course, the *Champaign-Urbana News Gazette* is a bit on the provincial side; any insect that doesn't eat corn or soybeans can pretty much give up hope of appearing on the front page in this town. But even the considerably hipper student paper, the *Daily Illini* didn't run the story, although one student did write an editorial column on the subject, called, "Mission insectible: Bush's bug-thugs strike back." I eventually found a version of the story in the *National Enquirer*. Look, I know what you're thinking, but you're wrong, I really don't read it all that often. I think my husband actually brought that issue home from the grocery store. Yes, that's it, my husband bought it and I just happened to see the story as I was about to recycle the paper. In any case, I was anxious to find a version of the story in the legitimate press, so I ended up going to the newspaper library on campus.

It seems that the *Washington Post* broke the story on 20 February 1990, with the headline "U.S. may try biological war on coca crop/Swarming caterpillars would devour plants." Clearly, this was the story I was after. After reading it, all I can say is that the editorial cartoonists didn't do it justice. It's not that I don't think the general principle was sound—the details were what struck me as amazing.

Take, for example, the statement in the *Washington Post* report that the object of all the attention, a white moth called the malumbia, "has not been written about in entomology journals for more than 55 years." Okay, so maybe lepidopteran systematists don't often pass through the Huallaga Valley of Peru and maybe drug lords don't routinely cooperate with cooperative extension agents, but, even so, I would have thought malumbias would have attracted someone's attention. So, I went back to the library in search of the malumbia, armed with the somewhat mangled scientific name *eloria-noyesi*, provided by the *Washington Post*. Rules for writing scientific names are simple and clear—capitalize the

How entomologists see insects

genus (the first part) and don't capitalize the species (the second part). I don't know why newspapers have such a hard time writing out Latin binomials, although I guess I shouldn't be surprised because no two papers seem to agree on how to spell Moammar Khadafi's name, either.

After exhaustive searching, I had to conclude that the *Washington Post* was right—I couldn't find any account of the malumbia in any entomology journals more recent than C.L. Collinette's 1950 revision of the genus. The closest I came to a study of insects attacking illicit plants was an article in the *Pan-Pacific Entomologist* about confused flour beetles infesting confiscated marijuana in a federal building in Douglas, Arizona. Actually, *Tribolium confusum* is called the "confused flour beetle" because it is frequently confused by entomologists with its morphologically similar congener *Tribolium castaneum*, the red flour beetle, but I expect these particular confused flour beetles may have been more confused than usual.

I did have better luck finding malumbias, however, in phytochemical journals. Murray Blum, Laurent Rivier, and Timothy Plowman published a paper in 1981 in the journal *Phytochemistry* describing the metabolism of cocaine by the elusive malumbia. Evidently, even though most of the ingested cocaine passes out with the frass, the caterpillar can sequester some of it from its host plant; female moths contain as much as 53 nanograms per gram body weight of the stuff. So here is an insect that actually has a street value. For that matter, here is insect excrement that has a street value. I can't imagine why these findings didn't get more publicity, and for that matter why legislation wasn't passed making it a felony to possess or smoke malumbia. This does raise a delicate etymological point, however—would the stub of a malumbia cigarette properly be called a roach?

Even more remarkable than the general lack of knowledge about the malumbia—after all, there are lots of tropical moths about which virtually nothing is known—was the fact that the Bush administration allocated $6.5 million dollars to this program. Granted, the money was for more than work on coca moths— there was also a project to test "a red dye that kills marijuana plants" (perhaps government stockpiles of banned Red 40 from maraschino cherries?) and to investigate "a soil fungus that wipes out coca." But, according to the article, "the principal focus of the stepped-up effort is the malumbia, a white moth that, in its caterpillar stage, gobbles the green leaves of the coca plant."

So basically, the government allocated more than six million dollars to breed and air-drop malumbias. That's six with six zeros after it. Six-oh-oh-oh-oh-oh-oh. That's a *lot* of money. Actually, that's three times the *entire* 1990 budget of the Plant Pest Program in the Competitive Grants Office of the U.S. Department of Agriculture. That's the program that funds research on caterpillars that gobble the green leaves of soybeans, corn, wheat, oats, peaches, pears, plums, turnips, cabbage, cauliflower, carrots, parsley, parsnips, celery, pine trees, cotton, tomato, potato, and other plants too numerous to mention.

I guess if there is a lesson here, it's that research funds are available to work on insects if you pick the right one. It has to be a species that clearly fits into a political agenda. Unfortunately, not too many insects fit this description. Mosquitoes bite conservatives and liberals alike, and termites cannot, as far as I know, be trained to eat up savings and loan buildings, kited checks, records of illegal campaign donations, copies of Walt Whitman's *Leaves of Grass*, or any of the other things that have compromised politicians of late (Dan Quayle's "potatoes" come to mind, too). I suppose it's best, then, that scientists focus on problems of scientific interest and remain objective, apolitical, and underfunded. I wonder, though—given former President George Bush's aversion to broccoli, was there possibly a secret fund during his administration to support studies of moths that, in their caterpillar stage, gobble those green leaves?

Pick a number from 1 to 10^{41}

I like to help people. Unfortunately, as an entomologist, I don't get a lot of people running to me for help. I guess if I were a doctor or a lawyer (or, for that matter, a police officer, telephone operator, librarian, auto mechanic, travel agent, or department store sales clerk), I'd get more questions. The discouraging thing about being an entomologist is that, frequently, I can't even help the people who do come to ask me questions. Usually, these kinds of questions begin with statements like, "I found this bug, and it's brown or black and I think it has six legs but I'm not sure." There is one question, though, that I'm ready for. Someday, someone will come into my office and ask me, "If all of the offspring of a single fly survived to reproduce, how many flies would there be in a year?"

I know the answer to this and similar vitally important questions, because entomologists have occupied themselves with making these vitally important calculations for years. The tradition goes back at least as far as J. H. Fabre, who did some figuring and reached the conclusion that "three flies will devour a dead horse as quickly as a lion." This fact, although titillating, must certainly have been a conversation-stopper at parties, especially the dead horse part, so it's not surprising that others felt compelled to improve on the estimates. Charles Darwin (1859) confined his geometrical increase estimates to vertebrates, calculating that one breeding pair of elephants would give rise to nineteen million progeny "after a period from 740 to 750 years. Nineteen million is indeed a large number, particularly in elephant units, but 750 years is also a long time, so even Darwin's calculation didn't impress everyone.

Insects were obviously the group of choice for generating staggering numbers without a wait. Following up on Fabre, Jordan and Kellogg (1908) estimated that "if each egg of the common house fly should develop, and each of the larvae should find the food and temperature it needed, with no loss and no destruction, the people of the city in which it happened would suffocate under the plague of flies." Daunting, yes, but hardly rigorously quantitative. L.O. Howard rectified that deficiency in his 1911 book *The House Fly—Disease Carrier.* He estimated that a single female fly who started to reproduce by 15 April in Washington, D.C., would have generated a population of 5,598,720,000,000 adults by September 10. As Howard justifiably remarks, "Such figures as these stagger the imagination."

That's probably why other entomologists felt obligated to improve on the calculation. Plowman and Dearden gently reminded readers in 1915 that, although Howard assumed that each fly lays only a single batch of eggs, "a fly may lay from 4 to 6 batches of eggs,...thus founding not one, but several, colonies in a single season." Hodge and Dawson (1918), not content with merely pointing out places for improvement, made their own calculations from scratch. They set their fly to laying eggs on 1 May (in an unspecified location), and estimated that, with 150 eggs laid at a clip (compared with Howard's 120), there would be 5,746,670,500 flies by 30 July—or, in more familiar units, "about 143,675 bushels of flies." Hodge and Dawson estimated the number of flies would escalate to 1,096,181,249,310,720,000,000,000,000 by the end of September. They leave their readers to convert that figure to bushels themselves. One must assume that it was Hodge, rather than Dawson, who had done the calculation, since he subsequently stated in another publication that, "A pair of flies beginning operations in April, might be progenitors, if all were to live, of 191,010,000,000,000,000,000 flies by August. Allowing one-eighth of a cubic inch to a fly, this number would cover the earth 47 feet deep."

Hodge would undoubtedly have been crushed to learn that Oldroyd (1964), an authority on flies, did not accept his calculations at face value: "Incredulous, I recalculated them and decided that a layer of such a thickness would cover only an area the size of Germany: but that is still a lot of flies." It seems unlikely that anyone will dispute the latter part of Oldroyd's conclusion any time soon. If nothing else, forty-seven feet of flies over every square inch of Germany would certainly wreak havoc with the tourism industry.

Insects other than house flies have attracted the notice of calculating entomologists. There's been a controversy raging in the literature almost as long as the fly furor about the reproductive capacity of aphids. Herrick (1926) took his lead from Huxley, who calculated that, after ten generations, the progeny of a single aphid "contain more ponderable substance than 500 millions of stout men; that is, more than the whole population of China." Herrick actually weighed four cabbage aphids (*Brevicoryne brassicae*), calculated the number of progeny from a single female after 16 generations (564,087,257,509,154,652), and estimated their collective weight at "789-odd quadrillion milligrams, which, by reduction, gives 789,722,160,512,816 grams" and by further and further reduction "gives us the staggering number of 822-odd million tons of ponderable substance." Figuring that the average stout man weighs two hundred pounds, Herrick concluded that Huxley's comparison with the whole population of China would be a gross underestimate, weighing "altogether a mere bagatelle of 50,000,000 tons."

Switching taxa, Howard (1931) reentered the insect fecundity fray by reconverting Herrick's 822 million tons back to pounds (1,644,000,000,000), estimating the average weight of a human at 150 pounds, the world population at 2,000,000,000, and the collective weight of humans on the planet at 300,000,000,000 pounds—"in other words, the plant-lice descended from one individual in a single season would weigh more than five times as much as all the people of the world." Calculating on the basis of length rather than weight, Metcalf and Flint (1928) decided "it would be possible, theoretically, for a single female to produce in one year, if all her descendents survived, a chain of these aphids long enough to encircle the earth," a far more robust estimate because the earth's circumference is less likely to vary than either the population of China or the average weight of stout men.

Other estimates of insect fecundity have failed to pass the test of time. Duncan and Pickwell (1939) cited the case of the vedalia ladybird beetle, *Rodolia cardinalis*—"if all circumstances were favorable to their survival, a population of twenty-two trillion beetles could be produced in six months' time! This is approximately twenty-two thousand times as many beetles as there have been minutes of time since the birth of Christ!" This conversion factor (as it were) doesn't really clear things up for me at all. Maybe another reason this particular calculation isn't frequently cited is that anyone who wanted to cite it would of necessity have to do some serious recalculations, since many minutes have passed since Duncan and Pickwell finished their arithmetic exercise. In fact, if it takes you a long time to do these calculations, you might have to start all over by the time you finish. For all I know, there are entomologists whose entire careers are subsumed by this task.

Or not. Nowadays, people don't seem so driven to come up with impressive numbers. Why do I say this? In 1954, Borror and DeLong introduced to this literature what may be the definitive calculation: starting with a pair of *Drosophila* fruit flies, allowing each female to produce one hundred eggs and allowing all progeny to survive, Borror and DeLong ended up, after twenty-five generations, with "about 10^{41}" flies. Just exactly how many flies is 10^{41} flies? "If this many flies were packed tightly together, 1000 to a cubic inch, they would form a ball extending nearly from the earth to the sun." I don't know about you, but I'm willing to take their word for it. Probably, most other people are, too. Significantly, in their first edition, Borror and DeLong prefaced their calculation with this statement: "Those who do not believe what follows may figure it out themselves." By the fourth edition, this statement is no longer included. After all, even if the ball would only reach as far as Mercury, that's still a lot of flies.

Ain't no bugs in me!

The human body comes equipped with nine or ten natural orifices, little portals that allow light, air, and solid or liquid material to enter or leave the body, depending on biological necessities. Although there are exceptions (which I'm sure, given time, you can probably come up with on your own), movement in or out of these orifices for any given state of matter tends to be resolutely unidirectional. Thus, it becomes rather unsettling whenever the normal flow of traffic is reversed. Drooling, for example, lacks the sensory fulfillment of drinking fine wine, and bleeding from the ears tends to be looked upon by most people with at least some degree of disquietude.

Unfortunately for us humans, insects are for the most part oblivious to these traffic patterns and thus occasionally wander into orifices that are not designed to accommodate them. Not all arthropods have an equal likelihood of appearing in any given orifice. Cockroaches, for example, appear to have a particular predilection for ears. According to one report published by Baker in 1987, of 134 foreign objects found in children's ears, 27 were arthropods, and, of these, 21 (78%) were cockroaches. In case you're wondering, one ant, one fly, three spiders, and a tick made up the rest. While there is a general consensus not only in the medical community but in the world at large that cockroaches do not belong in ears, there is by no means a similar consensus on the best procedure for removing said bodies from said orifices. The usual methods of dispatching insects are, for the most part, not really easily adapted to auditory canals—spraying insecticide directly into the ear seems only slightly less unpleasant than putting up with the cockroach, and dispatching the cockroach by stepping on it is just plain unworkable inside a person's head.

Physicians (as the experts to whom people who find cockroaches in their ears generally turn) have thus become amazingly resourceful. Among the most widely accepted approaches is to drown the cockroach lodged in the auditory canal in a fluid of some sort. A remarkable variety of substances have been used to this end, with varying degrees of success. While esoteric solutions involving benzocaine, succinyl choline, isopropyl alcohol, or hydrogen peroxide have been tried on occasion, the more prosaic water, vegetable oil, ether, and mineral oil have a long historical record of use. Of these, ether has the decided disadvantage of being explosively flammable and vegetable oil is rarely on hand in an emergency room. In 1980, Dr. A. Schittek introduced to an eager medical community a novel approach to the challenge of extricating cockroaches from auditory canals—immobilizing the cockroach with lidocaine spray. Lidocaine spray is more typically used as a topical anesthetic but when sprayed inside an infested ear it has the advantage of rendering a cockroach paralyzed and thus less likely to kick and scratch while being extricated.

This new method received validation of sorts when an unusual opportunity presented itself to an enterprising team of emergency room physicians in a large urban hospital; a patient checked in with a cockroach in *each* ear (O'Toole et al. 1985). The emergency team immediately set up a controlled study, using tried-and-true mineral oil in one ear and innovative 2% lidocaine spray in the other ear. While the cockroach that had drowned in mineral oil required manual extraction, the cockroach sprayed with lidocaine "exited the canal at a convulsive rate of speed and attempted to escape across the floor." In fairness, it must be pointed out that the "simple crush method," employed by a quick-thinking and "fleet-footed intern," was ultimately responsible for the demise of the cockroach, but the lidocaine clearly facilitated the process.

The field continued to advance in 1989, when Drs. J. Warren and L. Rotello improvised another method under very stressful circumstances. Although lidocaine was introduced into the

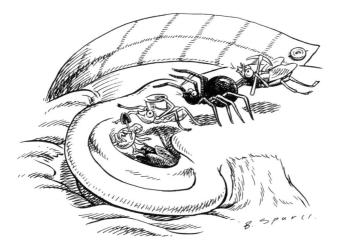

Orifice space for rent: immediate occupancy

auditory canal according to custom, it failed to have an instantaneous effect. Prompted by the patient's urgent request to "'Get that sucker outa my ear!'" the physicians took her at her word and applied a metal suction tip to the opening of the auditory canal; the cockroach was immediately sucked up and removed. These authors made medical history in that, in describing the moment of contact between cockroach and suction tip, they introduced the word "shloop" to the medical literature.

Although cockroaches appear to be the most frequently encountered insects in ears, the same cannot be said for other human orifices. Maggots have a habit of turning up in all kinds of openings, natural or otherwise. Maggots are what turned up, for example, in the urogenital tract of a 5-year-old girl in a Tokyo hospital. Some of these larvae came into the possession of R. Disney and H. Kurahashi (1978), who attempted to rear them to adulthood. Eventually, these authors tentatively identified the specimens as a species of *Megaselia*. Positive identification was undoubtedly complicated by the fact that, of the two larvae they were rearing, one "escaped"—although the authors do not describe how a legless, headless maggot encumbered by "very conspicuous posterior balloonlike structures" managed to make a clean getaway. Curiously, these authors made no attempt to speculate on how these maggots came to live where they did; in fact, there is no indication in the article that the authors thought that the habitat was in any way extraordinary, although they did allow as how they found the specimen "interesting."

Disney (1985) eventually described the specimen as a new species, *M. kurahashii*, having been supplied in the interim with additional specimens from one Dr. K. Kaneko, although where these additional specimens were collected was not specified. An earlier publication reports this species as breeding in steepers of Takuwan, a kind of Japanese "pickle made with radishes, rice-bran, and salt." The fact that the species initially found in a girl's urogenital tract also breeds in pickle brine doesn't really clear things up for me. For the life of me, I can't imagine any plausible scenario that connects Japanese pickle brine and urogenital tracts, but perhaps I am just lacking in imagination or suffering from too conventional an upbringing.

I guess I'm interested in how maggots in particular and insects in general gain access to human orifices because, as the possessor of more than a few of these orifices, I would like to take every precaution necessary to keep them insect-free. I've always believed that one of the few benefits of living in central Illinois is that one is relatively well insulated against the many forms of arthropod infestation that are largely limited to tropical climes. Human bot flies, jigger fleas, and Congo floor maggots are among the very few things I do not have to spend time worrying about on a daily basis. However, casual interloping at orifices that are left open and inviting seems to have no climatological boundaries. Badia and Lund (1994) describe a case of nasal myiasis, infestation of the nasal cavities, by *Oestrus ovis*, the sheep nasal bot fly, in a 35-year-old living in London, England. Nasal myiasis is not all that uncommon in tropical Asia and Africa—Sharma et al. (1989) reviewed some 250 cases over a ten-year period—nor is it all that uncommon in shepherds and in other people who for whatever reason choose to spend a lot of time around sheep. But this man from London denied having knowingly associated with sheep or traveled abroad immediately prior to the appearance of the maggots.

The mere occurrence of these maggots in the man's nasal

passages, however, was not the most remarkable thing about this case; what struck me as truly extraordinary was that this man had been "sneezing out several maggots during the preceding six weeks" before he checked in with his physician. Call me a wimp, but I think if I sneezed out even one little tiny maggot I would be on the phone and dialing 9-1-1 before it even hit the floor.

While it's true that London, England, is a comfortable 3,000 miles or so away, there's little justification for complacency here. M.J. Phelan and M. W. Johnson (1995) recently reported an instance of myiasis uncomfortably close to home. A 16-year-old boy returning from summer camp in southwestern Michigan experienced a rapid and progressive decline in the visual acuity of his right eye. Close examination of the eye revealed the presence of "a white, segmented maggot, approximately 1.25 disk diameters in length and tapered at both ends . . . moving slowly in the subretinal space near the equator inferotemporally." Lidocaine and mineral oil both being out of the question here (not to mention shoe leather), the inventive physicians photocoagulated the maggot with an argon laser, treatment end point being a "mild vaporization (bubbling) of the worm." Possibly the only thing more disconcerting than the thought of a maggot moving slowly across one's eye is the thought of a maggot being mildly vaporized while attempting to crawl across one's eye.

From even this superficial and incomplete review of a disturbingly vast literature, I have reached the inevitable and distressing conclusion that nobody's orifices are safe these days. I really don't mind that insects might occasionally take advantage of extraordinary circumstances. It's not all that surprising that debilitated geriatric patients in comas come to host infestations in their mouths, for example. And, although I couldn't actually read the paper (because it was in Japanese), the translated title of Tomita et al. (1984), including the words "self-amputation of the penis," suggests a set of circumstances that must certainly qualify as unusual by anyone's criteria (and not anything that I will have to worry about any time soon). But I never imagined my eyes, ears, nose, and mouth (not to mention less public places) might be at risk here in the Midwest. I don't know what to recommend—it's not as if we can go about our business with eyes shut tight and fingers in our ears. Maybe I'll think of something, but, until then, I can pass on one bit of advice—if you should find yourself in a Japanese restaurant, try to steer clear of the pickles when you sit down.

Getting up to speed

A while ago, I received a phone call from an editor at *Ranger Rick Magazine*, asking if I might verify a few facts for an insect story that was about to come out. This care and attention to accuracy came as no surprise to me—despite the fact that they're written for children, articles for *Ranger Rick* are scrupulously reviewed. I know this to be true because, the one and only time I ever wrote for the magazine—an article titled, "Watch out! Wild carrots!"—reviewers caught an error that went unnoticed in an article on a similar subject that went to a scientific journal for grown-ups. In particular, the editor wanted to know if the New Zealand weta (one of several species of very large stenopelmatid crickets) is heavier than the goliath beetle (one of several species of very large scarabaeid beetles). Frankly, all I knew at the time was that they were both really big insects, and, with intraspecific variation being what it is, providing a definitive answer could definitely be risky. I hesitated to go with my gut instinct and say "goliath beetle," without first ruling out the possibility that, lurking deep within the jungles of New Zealand, there might be a morbidly obese weta with a glandular condition. Moreover, I really didn't think it should matter to people whether average wetas are a fraction of an ounce heavier than average goliath beetles.

I know, though, as does the editor, that it really does matter. For reasons I can't completely understand, most people seem to care passionately about records. Students, for example, who complain about the burden of memorizing the names of insect orders can rattle off statistics about Chicago Bulls' superstar Michael Jordan's shooting percentage or Chicago Cubs rookie pitcher Kerry Woods' earned run average at will. They're willing not only to commit these numbers to memory but also to update them as they change (hey, it's not like the names of the orders change over the course of a semester). The world is awash in records and the most prominent keepers of records are the people at the Guinness Book of World Records (GBWR). First published in August, 1955, the book became a best seller within a matter of weeks and it has remained a best seller ever since; sales now approach $80 million a year.

The people behind the GBWR have not overlooked the class Insecta in their pursuit of all things exceptional or extraordinary. The book includes categories of achievement for which all animals are eligible—e.g., records for greatest concentration of animals (currently held by a swarm of *Melanoplus spretus* locusts sighted over Nebraska in July 1874 and estimated to have contained over 12.5 trillion insects), fastest reproduction (the cabbage aphid *Brevicoryne brassicae*), most acute sense of smell (the male emperor moth), the strongest animal (a rhinoceros beetle), and the most prodigious eater (larvae of the polyphemus moth). And there are records for which only insects are eligible—the oldest insect, the longest insect, the smallest insect, the lightest insect, the loudest insect, the insect with the fastest wingbeat, the insect with the slowest wingbeat, and so on. Here is where the Guinness people weigh in on the heaviest insect controversy, designating the goliath beetles (*Goliathus regius*, *G. meleagris*, *G. goliathus*, and *G. druryi*) as the collective record holders in the 1998 edition (although among coleopterists the taxonomic status of these four is in dispute, even if their size isn't). There are even a few records restricted to members of certain taxa—largest grasshopper, largest flea, longest flea jump, largest dragonfly, smallest dragonfly, largest butterfly, smallest moth, and longest butterfly migration.

"Fastest flying" is a category that's been around for a while and it's worthy of discussion because it illustrates the pitfalls of paying attention to these sorts of records. At the moment, according to GBWR, the record is held by *Austrophlebia costalis*, an Australian dragonfly clocked at 36 mph by person or persons unnamed. Historically, however, the zest for setting (or even just reporting) records has caused many people to lose their objectivity. The deer bot fly *Cephenemyia pratti* was assumed to be the fastest flyer on earth for a long time. *C. pratti* is one of a group of oestrid bot flies that make their living laying their eggs in the nostrils of deer and their relatives and developing as maggots by consuming blood and soft tissues in the nasal and pharyngeal cavities of their hosts. This insect became a record holder as a consequence of buzzing by one Charles Townsend as he was scaling 7,000-foot peaks in the Sierra Madres of western Chihuahua. The event apparently left an impression.

There was at the time an ongoing debate in the popular science literature, occasioned by new technologies in aeronautical engineering, on the feasibility of a daylight-day circuit of the earth. In an essay on the subject, Townsend (1927) put high speed travel in the context of his own personal experience: "the gravid females pass while on the search for hosts at a velocity of well over 300 yards per second—allowing a slight perception of color and form, but only a blurred glimpse. . . . On 12,000-foot summits in New Mexico I have seen pass me at an incredible velocity what were quite certainly the males of *Cephenemyia*. I could barely distinguish that something had passed—only a brownish blur in the air of about the right size for these lifes and without sense of form. As closely as I could estimate, their speed must have approximated 400 yards per second." For those keeping count, 400 yards per second is equivalent to 818 miles per hour (greater than Mach 2) and 300 yards per second is 614 miles per hour. Townsend reckons that these flies could likely have "kept up with the shells that the German big-bertha shot into Paris during the world war." Despite the extraordinary biological nature of this claim, Townsend really didn't seem all that impressed. Rather than dwelling on the astonishing nature of the fly's abilities, he devoted most of his essay to offering suggestions for inventing flying machines that can beat the speed of the earth's axial rotation, apparently a more easily realized goal than beating the speed of the fly.

Townsend may not have been impressed, but a lot of other people were. For over a decade, this record was cited widely— among other places, in the *New York Times* in 1926 and in the *Illustrated London News* in 1938—particularly to put into perspective feeble human attempts to set new speed records with mechanical devices. These citations eventually drew the attention of Irving Langmuir, an engineer with the General Electric Research Laboratory in Schenectady, New York. By use of dimensional reasoning, "comparing the fly with a Zeppelin as to diameter and speed and fuel consumption," along with ballistics equations and simple mathematics, Langmuir (1938) was able to calculate that a fly traveling at such speeds would have to consume 1.5 times its own weight in food every second in order to maintain itself. Moreover, flying at such speeds, a fly that strikes human skin "would come to rest in about 55×10^{-6} sec and during this time there would be a force of 1.4×10^{-8} dynes or 140 kg (310 pounds)," certainly enough force to "penetrate deeply into human flesh." Given that these flies have the habit of darting in and out of their host's noses to lay their eggs, it's remarkable that more Sierra Madre mule deer aren't wandering around with an extra nostril or two.

Based on the appearance of a moving lead weight on a string (and observing at what speed it becomes blurry), Langmuir estimated that Townsend's blurry flies were probably traveling only at the far-from-record-setting speed of about 25 mph. So, it's clear there's a need for stricter standards when it comes to reporting on record-setting animals. There are already strict standards for human accomplishments that involve insects, and I can see the value of reporting such records, even if they are a tad on the bizarre side. For years, there weren't many such records to worry about. In the 1998 issue of GBWR, in fact, among the "fantastic feats" documented (on the same page as the records for logrolling, ladderclimbing, bigamous marriages, and knitting) is, somewhat incongruously, the sole 1998 insect-related record—for wearing a mantle of bees. On June 29, 1991, Jed Shaner "was covered by a mantle of an estimated 343,000 bees weighing an aggregate of 80 pounds in Staunton, Virginia."

Television stands to propel insect-related human records out of their neglected status. Summer 1998 marked the debut of the Fox

Network television program, "Guinness World Records: Primetime." In early promotions, each episode was promised to include "multiple challenges and breathtaking events during which people go to the ultimate extremes" either to break existing records or create new ones. Some of these record-setting scenes are more visually appealing than others—walking a tightrope between two hot air balloons at 14,000 feet is probably more exciting to watch than, say, the man with the world's largest feet. The quest for ratings has added substantially to the number of insect-human records in the record book. In June, 1998, in front of GBWR judges, Dan Capps, by the act of spitting a dead cricket a distance of 32′ 1/2″, succeeded in setting a new world's record for dead-cricket spitting.

I spoke with Dan Capps about this feat when he came to the University of Illinois for the 1998 Insect Expo, accompanied by his remarkable collection of insect specimens. Mr. Capps was a 48-year-old maintenance mechanic at an Oscar Mayer bologna plant in Madison, Wisconsin when he accomplished this impressive feat. He was very self-effacing about his accomplishment and in fact revealed to me that the official record isn't even his personal best. On April 19, 1998, at Purdue University's Bug Bowl, Mr. Capps succeeded in spitting a dead cricket 32′ 1-1/2″ but, because the official judges were not present, that particular spit never made it into the record books.

There is no question that television has upped the ante in terms of the nature of records set; big risks mean big ratings. On October 20, 1998, Dr. Norman Gary, retired bee biologist from the University of California at Davis, traveled to Griffith Park in Los Angeles, CA, and, in front of officials, succeeded in holding 109 live bees in his mouth for ten seconds, thereby setting a world's record for holding live bees in the mouth for ten seconds. This world record was one that Dr. Gary conceived of himself, based on years of working with bees and studying their behavior. I won't reveal to you his secrets; suffice it to say that, even though "Guinness Primetime" compensates its record-setters, there isn't enough money in the world to entice me to attempt to break this one.

For the record, Dr. Gary is no stranger to record books; he says he's very competitive by nature. His first record, and first encounter with GBWR, was back in 1988, when he set the Australian record for largest mantle of bees. In 1998, he conquered the world—on July 21, 1998, Dr. Gary succeeded in assembling a mantle of bees on colleague Mark Biancaniello, an animal trainer who had worked at Michael Jackson's Neverland ranch, that weighed in excess of 87.5 pounds and included an estimated 353,150 bees. It took three tries, and it required developing new and innovative methods for estimating the number of bees in a mantle, but Dr. Gary rose to the challenge and earned a form of immortality in the process (at least until someone comes up with an 88-pound bee mantle).

I'm happy for Dr. Gary, but I don't want to be the one to break the news to Jed Shaner that he will no longer be featured in GBWR in the "fantastic feats" section—at least for "mantle of bees." If he's been busy ladder-climbing, knitting, or logrolling since 1991, he may still have a shot in the next issue on the same page. And I'd encourage those wetas not to lose hope—there may be a place for them in the next issue if they keep eating.

Sea monkey see, sea monkey do

One of the great disappointments of my childhood was the fact that my parents never allowed us to have any real pets. By "real pet," I mean any creature capable of learning its own name. All of the goldfish, red-eared turtles, and anoles we were allowed to keep, then, didn't really count. Nor did Jacques, our hamster, the only mammalian pet to grace our home during my formative years. I know for a fact Jacques didn't answer to his name, because, when he managed to escape from his cage one fateful day, we called out his name over and over again in the hope that he would materialize, and he never did. Months later, my mother found his tiny, shriveled body in a remote corner of the attic. To this day, I feel bad about Jacques.

I think my parents were reluctant to allow us to have pets because they were concerned that we weren't responsible enough to take care of a sentient creature (and I guess that unfortunate hamster incident pretty much proved their point). Nonetheless, even in my seriously pet-deprived state, I never had any interest in owning an insect pet. This is all the more surprising given that I grew up during the dawn of the insect pet era. On July 4, 1956, Milton Levine poured some sand into a plastic container and invented the Ant Farm. Uncle Milton, as he and his eponymous ant farms came to be known, really found a market niche; today, it's a million-dollar enterprise, with close to seven million ants sold to ant farm owners annually. Personally, though, I've never been tempted by the prospect of ant farm ownership. Knowing as I do today that the species of choice for populating these farms are *Pogonomyrmex* harvester ants, which have among the nastiest stings in the class Insecta, I can't help wondering what percentage of Uncle Milton's profits go toward maintaining a crackerjack legal staff.

Had I been less adamant about that name thing, I could even have gotten in on the ground floor of the sea monkey phenomenon. Sea monkeys are the arthropod pets, par excellence. Only a year after ant farms came into existence, one Harold von Braunhut, of Long Island, New York, had a brilliant flash of insight relating to marine biology and its commercial pet potential. Three years later, in 1960, he was marketing genetically improved *Artemia salina* brine shrimp (hitherto vended as fish food) as "Instant Life"—pets that came to life simply by the addition of water. As a child, I wasn't tempted by sea monkeys any more than I was by ant farms. Among other things, they were always advertised in the back pages of comic books next to the ads for X-ray specs, and, even at that early stage of my scientific training, I was fairly certain that X-ray specs couldn't possibly work as advertised. I also had a hard time believing that simply adding water could reanimate a living creature; instant oatmeal I could accept, but not instant animal.

I was wrong, however, and sea monkeys went on to become an international phenomenon. Today, countless children add water and enjoy their instant pets; they achieved such popularity that they even inspired a short-lived television show, on Saturday mornings on CBS, in the early 1990s. According to the catalogue and instruction book that accompany every Sea-Monkey Ocean Zoo, dozens of products are available for the sea monkey owner who wishes to indulge his pets. There's an Electric Ocean-Zoo "Showboat," a "Sea Show Projector," "Sea Medic Sea-Monkey Medicine," and even "Sea-Monkey Banana Treats," to reward sea-monkeys "for the FUN they give you!" Not surprisingly, the book provides tips on feeding and breeding sea monkeys, but it also has a section on training sea monkeys to perform tricks and to play games with people. For an extra $1.25, there's a supplemental book to teach them how to play baseball (according to patent number 3,853,317 issued to the redoubtable von Braunhut).

The last page of the instruction book provides a "limited group sea-monkey life insurance policy," with a form on which to write the names of sea monkey pets. Also provided is a naming guide:

> "Names given must be Socially Acceptable, i.e., names such as : Stinky, Slimy, Sneaky, etc. will not be allowed as your sensitive pets might be offended. Give them nice "Sunday School" names. Suggestions: Scamper, Moby Dick, Davy Jones, Barry Cuda, Barry Goldwater, Sharkey, Agamemnon, Puddles, Finn, Peppy, Flippy, etc."

I notice, though, that nowhere is there any kind of guarantee that they'll answer to those names.

I know it shouldn't, but it bothers me, as a professional entomologist, that one of the "World's Most UNUSUAL and AMAZING Pets" is a crustacean and not an insect. I know today that von Braunhut created a novel pet by taking advantage of the phenomenon of cryptobiosis, or anhydrobiosis—kind of a suspended animation state induced by desiccation. The phenomenon has been reported in a wide variety of crustaceans other than brine shrimp, including ostracods and water fleas, as well as nematodes and tardigrades, but remarkably few insects are very proficient at entering a cryptobiotic state. Among the few exceptions, and perhaps the best known cryptobiotic insect, is *Polypedilum vanderplanki*, brought to the attention of the scientific world by H. E. Hinton in 1951. That it was H. E. Hinton who brought *P. vanderplanki* to prominence is not really surprising; throughout his long and productive career, H. E. Hinton brought all kinds of remarkable things to the attention of the scientific world, including lycaenid butterfly pupae that look like monkey heads. *P. vanderplanki* is a chironomid midge that, as a larva, lives in small pools that form during the rainy season in depressions in unshaded rocks in Nigeria and Uganda. When these pools dry up, the larvae dry up with them, often while ensconced in burrows made in the thin layer of mud that lines the bottom of the pool. Pools can fill up and dry out alternately several times during the rainy season and during the dry season temperatures at the surface of the dry soil layer protecting the larvae can exceed 42°C.

Intrigued by the challenge, Hinton commenced a decade-long effort aimed at determining the physiological limits of *P. vanderplanki*. In 1951, he reported that, at 0% relative humidity, the water content of the larvae decreased to about 3%; in the laboratory, larvae could tolerate up to ten successive dehydrations and rehydrations without ill effects. Two years later, Hinton reported that desiccated larvae maintained at room temperature and humidity could survive for more than three years and still be reanimated simply by rehydration without adverse effects. Storage for three years at room humidity followed by seven years of storage over calcium chloride also failed to prevent larvae from reanimating upon rehydration (Hinton 1960).

Hinton continued to push *P. vanderplanki's* envelope, subjecting them to even more rigorous conditions. He discovered that they could survive exposure to 106°C for 3 hours and 200°C for five minutes. And they were unfazed by total immersion in absolute alcohol for seven days, in glycerol for 67 hours, in liquid air, at −190°C, for 77 hours and in liquid helium, at −270°C, for 5 minutes.

I don't know about you, but I think surviving total immersion in liquid helium is a pretty cool trick (as it were), much more impressive than, say, Sea Monkey hypnosis (positive phototaxis) or Sea Monkey Acrobatics (swimming). Yet, *P. vanderplanki* larvae have never lived up to their obvious pet potential. Perhaps all they're lacking is a catchy common name. Tardigrades, or "water bears," may already have the edge on them in that regard. Moreover, in their cryptobiotic state, water bears can not only withstand temperatures as low as −253° and as high as 151° but can survive a century of desiccation as well as exposure to a vacuum, to X-rays, and to hydrostatic pressure equal to six times the pressure of sea water at 10,000 meters depth (Seki and Toyoshima 1998).

But maybe it's just as well; having children plunge their pets in liquid helium doesn't really seem like the right mechanism for teaching responsibility. As a parent myself now, I have to think of these things. Maybe I'll just order my daughter a pair of X-ray specs and hope for the best.

A prayer before dining

If you stop almost any average citizen on the street and ask him or her to provide you with three facts about insects, odds are good that one will be this: the female praying mantis is a cannibal that is not beneath eating her own mates or children. Absolutely everyone seems to know this particular bit of insect lore and it's practically celebrated in popular culture. It's been featured in "The Far Side" cartoons ("I don't know what you're insinuating, Jane, but I haven't seen your Harold all day—besides, surely you know I would only devour my *own* husband!") (Larson 1987) and it's even provided the plot for at least one episode of "Buffy the Vampire Slayer" on television—the one called "Teacher's Pet," which, according to *TV Guide*, features a male high school student "nearly seduced by a voluptuous substitute science teacher who transforms into a large praying mantis . . . [And] what's more embarrassing than almost getting devoured by a femme fatale insect teacher?" It has even made it into the screenplay of a James Bond film. In Dr. No (1962), voluptuous Honey Ryder (who later murders a man by placing a black widow spider under his mosquito netting) says, "Did you ever see a mongoose dance or a scorpion with sunstroke sting itself to death, or a praying mantis eat her husband after making love? Well, I have." Even people who can't keep straight in their minds the concept that spiders aren't insects seem comfortably fluent with the notion that praying mantids are unreconstructed sexual cannibals.

While it's true that consuming offspring is fairly widespread in the animal kingdom, as anyone who has tried to raise gouramis in a fish tank that's too small can attest, the sexual cannibalism thing is a source of particular fascination. Mantids are by no means the only arthropods that are reputed to engage in the practice—spiders, scorpions, amphipods, copepods, crickets, grasshoppers, antlions, and ground beetles are known to indulge from time to time. But mantids seem to hold a special place in the pantheon of sexual cannibals. After all, it's not pictures of cannibal copepods you see in the introductory biology textbooks. Figure 55–14b in Helena Curtis' *Biology* (p. 1032), for example, depicts "copulating praying mantids. This male mantid is lucky—so far. Female mantids usually eat their mates, often decapitating them before copulation. Decapitation of the male mantid releases inhibition and results in his copulating even more vigorously, thus helping to ensure that his sacrifice will not have been in vain." The implication is that the male's fate, however happy it might be in the short term, is pretty much sealed permanently.

The only problem I have with this venerable fact of life is that it's not at all clear to me that it's a fact at all. Among other things, there are more than 180 species of mantids, and sexual cannibalism has been reported to occur in just a tiny handful of those species. Moreover, the vast majority of those reports are from laboratory studies, which are, needless to say, conducted under highly artificial conditions. Examining those reports in detail is quite interesting, entirely independent of their content. The paper that catapulted mantid sexual cannibalism into the American scientific conscience was the lurid account published by L. O. Howard in *Science* in 1886. It's only 500 words long but it makes up in impact what it lacks in verbosity:

> . . . I brought a male of *Mantis carolina* to a friend who had been keeping a solitary female as a pet. Placing them in the same jar, the male, in alarm, endeavored to escape. In a few minutes, the female succeeded in grasping him. She first bit off his left tarsus, and consumed the tibia and femur. Next she gnawed out his left eye. At this the male seemed to realize his proximity to one of the opposite sex, and began to make vain endeavors to mate. The female next ate up his right front leg, and then entirely decapitated him, devouring his head and gnawing into his thorax. Not until she had eaten all of his thorax except 3 millimeters did she stop to rest. All this while the male had continued his vain attempts to obtain entrance at the valvules, and he now succeeded, as she voluntarily spread the parts open, and union took place.

To me, even more remarkable than the phenomenon of sexual cannibalism is the fact that, back in 1886, you could get a paper published in *Science* based on a study with a sample size of one. In any case, Howard was evidently so fascinated with the phenomenon that he managed, with C. V. Riley, to publish a second account, this time based not on observation but on an anecdote related by one Colonel John Bowles about a captive pair he had observed. The fact that Bowles chloroformed the couple before the male could finish mating and the female could finish eating makes interpretation a bit difficult and is probably the reason this paper was published in *Insect Life* and not *Science*. The story reached the general public when masterful writer J. H. Fabre gamely picked up the sexual perversion gauntlet and in 1897 wrote a flowery, even moving, account of yet another captive male's spirited demise: "if the poor fellow is loved by his lady as the vivifier of her ovaries, he is also loved as a piece of highly flavored game. . . . I have . . . seen one and the same mantis use up seven males. She takes them all to her bosom and makes them pay for the nuptial ecstasy with their lives."

The paper that absolutely guaranteed "textbook example" status for mantid mating habits was the extensive study published by physiologist Kenneth Roeder in 1935. Roeder is widely credited with suggesting that sexual cannibalism is required among mantids because inhibitory impulses from the subesophageal ganglion prevent the mantis from completing his conjugal duties; removal of the head removes these inhibitions and allow consummation to take place. I doubt, though, that many people have actually read this paper. Roeder didn't really go so far as to suggest that decapitation was a necessity. Among other things, he was well aware that, in nature, many mantids mate multiple times, and he was the first to admit that the conditions under which he made his observations were, to say the least, artificial.

I doubt, though, that reading Roeder's clarification would convince people to abandon the notion that mantids eat their mates. Later studies failing to document cannibalism at all (Liske and Davis, 1984) or documenting cannibalism only under certain ecological circumstances or at levels well below those meriting the statement "Female mantids usually eat their mates" (Lawrence 1992) certainly haven't. People hate to let go of things sick and twisted; after all, there's great reluctance to let go of the notion of human cannibalism, despite the fact that the evidence for it is flabby, indeed. According to Brottman (1998), in his *Meat is Murder! An Illustrated Guide to Cannibal Culture* (note: this is *definitely* not recommended as a coffee table picture book), "the major historical phenomenon is the idea that people eat each other, not the fact." Since time immemorial, it has been the practice to note that "the other fellows" are cannibals. Herodotus, widely recognized as the first anthropologist, described "Androphagi," with "the most savage customs of all men," in the eastern fringes of Europe. Throughout history, Romans accused Christians, Christians accused Jews, the English accused the Scots and Picts, and Europeans accused Africans, New Guineas, Polynesians, Native Americans, and just about any other non-European people they encountered. The cultures most likely to display cannibalistic traits have, not coincidentally, tended to be the ones in possession of material goods or resources most highly desired by the reporters of the cannibalism. There very well may be occasional incidences of cannibalism; historically, however, it's highly likely that the accusation takes hold as a mechanism, acknowledged or not, to marginalize a people and to justify subsequent acts of violence against them. In fact, the word "cannibal" itself is a reference to the "Caribe" people, who resisted pacification efforts by Columbus and his successors; the Arawak, a more tractable tribe in the same neighborhood, were never accused of such dietary anomalies.

After all, there is ample documentation that cannibalism exists in western societies. Brottman (1998) provided graphic evidence of that fact, with stories of Fritz Haarman, the "Butcher of Hannover," a meat-vendor who not only ate vagrant children but sold their flesh as horsemeat in his butcher shop in Hannover, Germany; Anna Zimmerman, of Monchen-Gladbach, Germany, who killed her lover and then chopped him up into "manageable pan-sized steaks;" Karl Denke, the "Cannibal Landlord" of Munsterberg; "Weird Old Eddie" Gein of Plainfield, Wisconsin (the inspiration for the movies "Texas Chainsaw Massacre," "Psycho" and "Deranged"); and, of course, Jeffrey Dahmer, the "Milwaukee Cannibal." But nobody would seriously suggest that humans are, as a species, cannibalistic, or even, from the face of it, that people from Germany or Wisconsin are cannibals.

Just as people would like to believe the worst about another culture due to be subjugated, I think people would like to believe the worst about insects, an entire class that most people would like to subjugate. I think that's one reason the tortuous hypothesis of adaptive mantid cannibalism still remains firmly entrenched in the scientific literature. There are, after all, other explanations for the fact that, when the subesophageal ganglion is cut, the male genitalia start pumping away. One that comes to mind is nonadaptive inhibition—it's in the nature of the wiring. Well known to neurobiologists, such release phenomena are described as "abnormal responses to stimuli or of motor behaviors that emerge" after damage to the corticospinal system (Kandel and Schwartz, 1985). These abnormal responses are generally attributed to the removal of inhibitory signals that influence the interneuronal networks controlling the response. Roeder (1936) himself reported that the genitalia of decapitated *female* mantids also come to life—but no one ever suggested that any female needs to lose her head before she engages in sexual intercourse.

Inhibitory impulses are well documented and, when strange things ensue after their removal, most people are rarely moved to erect elaborate hypotheses to account for them. Here's a case in point—Schmidt et al. (1999) studied penile erections during paradoxical sleep in rats and humans. It's been long suspected that the brain exerts an inhibitory signal to the organ because (I'm quoting here) "reflexive erections are facilitated by spinal transections or spinal block." All kinds of lesions to the brain seem to release reflexive erective activity in rats—transection of the brainstem caudal to the medullar paragigantocellular nucleus, bilateral cytotoxic lesions to the medullar paragigantocellular nucleus, even complete midthoracic spinal transections not only don't stop reflexive erections, in some cases they even make them "more easily elicited with shorter latencies relative to controls." In other words, doing some selective brain surgery would do wonders for how some men may perform sexually (a speculation no doubt made independently by many women, even those unaware of these studies).

Needless to say, nobody is translating these results into tips for marital aid manuals, nor will bilateral cytotoxic lesions to the medullar paragigantocellular nucleus be replacing Viagra any time soon. It just wouldn't be reasonable. I'm not convinced it's so reasonable for mantids, either. Reports of sexual cannibalism seem better suited for the movies or maybe a German cookbook than for introductory biology texts.

Grotto glow

I've only been to the state of Arkansas once in my life—I spent the bulk of summer 1974 in Fayetteville—but that one visit has had a lifelong impact on me. I don't mean the fact that I now have to cart around a red two-pound candle shaped like a University of Arkansas razorback hog every time I change residences (I'm still not sure why I bought it back then and I really don't know why I've kept it all this time). Rather, I mean the raging claustrophobia that I contracted while I was there. I ended up in Fayetteville that summer as a student member of a research team charged with conducting an ecological inventory of Devil's Den State Park. Devil's Den, in the Boston Mountain section of the Ozarks in the state's northwest corner, was of ecological interest because it lay directly in the path of a proposed expansion of Interstate 71. Our team was supposed to inventory the animal and plant life in the park, paying particular attention to whether any rare or endangered species might be in residence. I was designated the team's invertebrate biologist despite the fact that the sum total of my experience consisted of exactly one course in terrestrial arthropod biology and one semester of a two-semester sequence in invertebrate zoology.

BUZZWORDS

There was yet another reason I was not exactly prepared for the assignment. Devil's Den owes its name to the extensive cave system that runs through the park and of course the caves were to be a central focus of our inventory efforts; caves have long been known to harbor strange and unusual life forms that can potentially stop highway projects. I'd actually never set foot in a cave before my trip to Arkansas, so to say I was speleologically challenged is an understatement. The names of the caves we were to explore didn't exactly inspire confidence. The state park owes its name to the local legend that early settlers heard 'the roar of the devil' in the vicinity; the two major formations in the park were called Devil's Den and Devil's Icebox. For the record, in addition to a Den and an Icebox, the Devil keeps a Kitchen, a Kettle, a Fireplace, a Dining Table, a Punch Bowl, a Sugar Bowl, a Honeycomb, and a Well in Arkansas; his Toll Booth is apparently somewhere north in Missouri.

It was on my first trip inside one of what are so aptly called crevice caves that I discovered I really can't cope with pitch blackness or narrow spaces you can't stand up or turn around in. Since no one else on the team seemed to be concerned that we might be buried alive at any minute, I managed to keep my feelings to myself. I struggled through the entire summer, though, desperately trying to fight back the blind panic I experienced every time we entered anything resembling a cave. There are evidently some unique biological features of the cave system in Devil's Den state park. There is, for example, an overwintering site (hibernaculum) for the endangered Ozark big-eared bat (*Plecotus townsendii ingens*), among the rarest bats in North America. I don't recall ever seeing any Ozark big-eared bats. I found some kind of crustacean once in the cave, which I couldn't identify (I suppose I really should have taken the second semester of invertebrate zoology, after all), along with quite a few Polaroid film wrappers and some empty beer cans, but otherwise I really didn't do much to expand the body of knowledge of Ozark cave biology. So I didn't have much of an impact on Arkansas' environment. But the Arkansas environment had had a definite impact on me—by the end of the summer, I was so claustrophobic that I couldn't walk into the elevator in the high-rise dorm where we stayed at the University of Arkansas campus.

Devil's Den State Park is at least part of the reason that, when I traveled to Australia in 1999, I didn't avail myself of one of the more unusual ecotourism opportunities in the world, an opportunity that certainly should have appealed to me as an entomologist. I did hear a talk about it, though, at the 1999 Australian National Congress. The talk, given by Claire Baker and David Merritt of the Department of Zoology and Entomology at University of Queensland, detailed the tourist industry geared around *Arachnocampa flava*, a cave-dwelling fungus gnat maggot that glows in the dark. *A. flava* lives in caves in the rainforests of southeastern Queensland. The immature stages of *A. flava* spin sticky threads that hang down like fishing lines; the bright blue-green glow of the larvae apparently attracts small prey, which get ensnared in the lines and become paralyzed upon contact with oxalic acid droplets distributed strategically along the lines. The maggot then hauls in the prey and consumes it.

There are a few other spots for viewing glow-in-the-dark maggots throughout Australia. There's the Glowworm Tunnel in Lithgow, New South Wales, for example, where luminous maggots light the ceiling of an abandoned railway line through the Blue Mountains constructed for oil shale workers. But the real mecca for watching fungus gnat maggots glow in the dark is in neighboring New Zealand, in the Waitomo Caves. Up to 400,000 tourists a year pay $20 (NZ) apiece to travel by boat through the Glowworm Grotto to see the luminescent larval, pupal, and adult *A. luminosa*. The glow is produced by modified Malpighian tubules in the last abdominal segment, which lie directly over a richly tracheated reflective layer. The fungus gnats apparently put on quite a show, glowing more brightly when fighting amongst themselves as maggots or when courting and mating.

The tourist industry discovered the Glowworm Grotto just about the same time that the scientific community became aware of the glowworms therein. The cave was first explored in 1887 by a local Maori chief, Tane Tinorau, and a British companion, Fred Mace, who instantly saw its commercial potential. By 1910, a hotel was built to accommodate the crowds of visitors. The entomological community first heard about the insects in an article published by 1886 in *Entomologists' Monthly Magazine*. Meyrich reported finding large numbers of sticky, luminous larvae along a steep creek bank near Auckland producing light consisting "of a small, bright, greenish-white, erect flame, rising from the back of the neck." Although he guessed that they were predaceous, and possibly beetles, he was loathe to put a name to them, claiming that "it is impossible for a wandering entomologist to attack a larva of these habits." Subsequent contributors to *Entomologists' Monthly Magazine* undertook the task, Hudson (1886) in the process pointing out that Meyrich's erect flame rising from the back of the neck was more like a brilliant gleam arising from "the *posterior* extremity of the larva," an understandable discrepancy given the general absence of heads and other distinctive directional features displayed by maggots. Osten-

Sacken (1886) was the one who eventually recognized it as a mycetophilid fungus gnat and even in a subsequent paper offered free copies of his recently reprinted review of larvae of Mycetophilidae to "anyone applying . . . for them."

Remarkably, given that there are only about a dozen species of luminous mycetophilids in the entire world, there are a few glow-in-the-dark species right here in North America, some of which aren't even too far from Arkansas. *Orfelia (= Platyura) fultoni* is a bluish maggot that is found in permanently damp soil in rock crevices or rotten wood in parts of the southeastern U.S. There's even a small tourist industry just beginning in Alabama, where visitors are invited to come to Dismals Canyon to see the "dismalites" in the glowworm-covered mossy canyon just down Highway 8 from the town of Phil Campbell. If the state of Alabama ever needs an environmental impact statement on improving Highway 8, I might even volunteer. I would feel a lot better about going into a cave or cavern there to look for insects knowing they lit up the place.

Not everybody, though, is as comforted as I am by the soft glow of arthropod light. In parts of Thailand, according to Yuswasdi (1950), many rural people believe "that a certain luminous myriapod, usually found in old thatched roofs, and known in Siamese as Maeng-Kah-Reaung ("luminous insect in the roof") has the habit of climbing into the ear of sleeping individuals to bore its way into the brain, where it prefers to dwell. Patients frequently complain of such intrusion, though no one seems to have actually seen the creature inside the ear. The chief complaint is an intermittent or continuous ringing in the ear of long duration." I'm not sure I believe these reports—after all, a physician wouldn't even need an otoscope to see a glowworm in a patient's ear so the fact that they haven't yet been spotted leaves room for skepticism. Moreover, there are two genera of luminescent millipedes and the genus found outside Asia, *Motyxia*, is reported to occur in the mountain valleys of California, where there haven't been any otherwise inexplicable outbreaks of ringing of the ears. On the other hand, maybe I should keep my phobias down to a manageable number and just try to stay out of thatched houses in California mountain valleys from now on.

How
the world
sees
insects

Super systematics

One of the nice things about living with a child is that one is afforded opportunities to become acquainted with icons of popular culture that might otherwise be overlooked. I have, for example, become quite conversant with characters that appear on Saturday morning cartoons. Not only can I sing the "My Little Ponies" theme song in its entirety, I can also recognize and name all of the Rugrats (including nonrecurring characters) and can distinguish between Yakko and Wakko Warner on "Animaniacs" (as can, needless to say, my daughter Hannah). But even my extensive Saturday morning experience didn't prepare me for one particular animated character—"The Tick," the title character on a short-lived series on Fox network. The Tick appears to be a tall man in a blue skintight outfit with two antennae dangling limply from the hood. Baffled by what I saw, I consulted my husband, Richard, who, as then-president of the Society for Animation Studies, certainly qualified as an expert on animated cartoon character. Richard attained this lofty status in life despite the childhood trauma of discovering that his mother had thrown away his entire comic book collection, including a number of mint condition vintage issues of "Uncle Scrooge." Richard informed me that the

Tick came to television after a limited run on the printed pages of comic books.

After calling a few comic book stores in town, I did locate one, A-Plus Comics and Sports Cards Shop, which carried back issues and re-releases of The Tick. According to *The Tick's Giant Circus of the Mighty* (Edlund, 1992), the Tick's alter ego, Neville Nedd, is the Weekly World Planet crossword puzzle editor. His superhero origins are obscure—evidently suffering from amnesia, his first memory is of escaping from a mental asylum called the Evanston Clinic. He professes to be a "blood-sucking arachnid" but has never been observed living up to his reputation and actually consuming blood or any other kind of body fluid. He is very strong and has the usual assortment of superhero gadgets, including a Hypnotic Secret Identity Tie, a Secret Crime View Finder, the Mighty Diner Straw, and the Pez Dispenser of Graveness.

All told, the Tick didn't strike me as a very impressive superhero. In the hope that other arthropods have served as the inspiration for more remarkable superheroes, I consulted *The Encyclopedia of Superheroes* (Rovin, 1985). In his preface, Rovin writes ". . . one can't classify superheroes with the finicky detachment of an entomologist distinguishing between varieties of insects." As a finicky entomologist, though, I couldn't help but feel that much could be gained by just such a classificatory scheme. For the most part, arthropod-based superheroes are easily placed in well-defined taxa. Arachnids far outnumber insects and include in their ranks the Scarlet Scorpion, the Scorpion, the Spider Queen, the Black Spider, the Black Widow, Spider, Spider-Man, Spider Widow, Spider Woman, the Tarantula, and the Web Queen. Running a close second to the arachnids are hymenopterans: Ant Boy, Ant Man, the Green Hornet, the Queen Bee, the Red Bee, the Wasp,

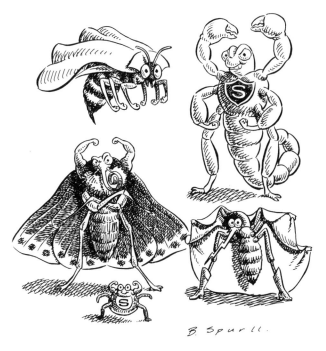

Members of the Superclass Insecta

and Yellow Jacket. The Coleoptera are well represented by Blue Beetle, Firefly, the Silver Scarab, and the Blazing Scarab, and the Lepidoptera by the Butterfly, the Moth, Gypsy Moth, and Mothman. Among the minor orders (at least among comic book arthropods) are Diptera (the Fly, Mosquito Boy), Odonata (the Dragonfly), and Hemiptera (Ambush Bug).

Certain aspects of arthropod biology appear repeatedly in the pages of comic books irrespective of the superhero's taxonomic status. Most of these superheroes have the "proportionate

BUZZWORDS

strength" of insects—the familiar old misinterpretation of the surface-area-to-volume ratio. Insects appear to have disproportionate strength because their surface area is large relative to their volume; muscle strength is proportionate to cross-sectional area so insect muscle, which moves very little volume relative to what human muscle has to move, appears to be quite strong. In the real world, an arthropod the size of a human would possess the relatively unimpressive strength of a human. In the comic book world, though, Spider-Man's alter ego, Peter Parker, "bitten by a radioactive spider . . . has gained that insect's proportionate strength." In the comic book world, too, spiders, which are of course really arachnids, are considered insects. The Fly possesses "muscles 100 times more powerful than humans." Ant Man presents an interesting variant on the theme—he is capable of shrinking to the size of an ant while retaining his human strength. He also possesses the ability to communicate with insects and order them to do his bidding.

Another recurrent theme is the ability to deliver venom, or at least an electric equivalent thereof. Yellow Jacket can shoot "energy stingers," the Scorpion can shoot " bug tracers," the Red Bee has a "stinger gun," Spider Woman fires "bioelectric venom blasts," and the Wasp possesses "sting wristbands." Even Mosquito Boy can sting, something his arthropod equivalents can't do. Many of the arthropod superheroes can scale buildings with the assistance of a combination of suction-cup devices on their feet (Tarantula, Black Widow, Butterfly) and a resolute disbelief in the laws of gravity. In some cases, attributes are highly taxon-specific. Web-shooting devices are restricted to spider-based superheroes— the Black Widow has her "widow's line," Tarantula a "web gun," Web Woman her "web rope," and Spider-Man a "web shooter."

How the world sees insects

Spider-Man 2099, a genetics engineer of the future, has the genetic code of a spider accidentally imprinted on his own DNA, conferring upon him most spider attributes, except for the ability to "shoot webbing out of his butt"—instead (perhaps in the interest of decency), it comes out of spinnerets on his forearms.

What bothers me most about this assortment of arthropod superheroes is not so much the liberties taken with arthropod biology but the human dimensions of these characters. Every superhero, arthropod or otherwise, has a story to explain how his or her superpowers came to be. Over the years, many superheroes began their careers as scientists, although by far and away the most common profession for superhero alter egos is millionaire playboy, this being the only occupation that permits you to disappear for days at a time saving the world without having to explain your absence to your boss. Some are generic scientists—sort of computer scientist/roboticist/engineer/nuclear physicists—but others are highly specialized professionals. There are oceanographers (Amphibian, Piranha, Stingray), biochemists (Beast, Giant Man, Steel the Indestructible Man), zoologists (Bwana Beast, Jaguar), geophysicists (Havak, Polaris), and, most of all, physicists (The Atom, Captain Britain, Dr. Fate, Dr. Solar, The Hulk, Human Bomb, Mr. Fantastic, Sasquatch, and Static). Distressingly conspicuous by their absence in this bunch are entomologists.

It seems that arthropod superheroes almost never owe their origins to the scientific insights of an entomologist. Whereas chemists can develop chemicals to enhance strength, and physicists can manipulate atomic forces to allow themselves to defy gravity, entomologists as a group appear incapable of applying their scholarship to save the world from crime and evildoers. More likely than not, arthropod superheroes owe their origins to accidents. Insect Queen, for example, alias Lana Lang, newscaster, possesses a Biogenetic Ring that allows her to assume the form of any arthropod; the ring was a gift from a six-armed alien whom she rescued from underneath a fallen tree. Ambush Bug stumbled across an alien space suit that conferred teleportational powers upon its wearer; thus can Ambush Bug ambush his adversaries. Thomas Troy became the Fly after finding a fly-shaped magical ring; when he's not busy crawling up walls, he's a lawyer. If that weren't bad enough, the 1975 version of Tarantula is an investment counselor when he is not doing super-deeds. Although not an entomologist, Ant Boy at least acquired his powers honestly, having been raised from infancy by ants. About the only entomologist in the whole bunch was a character called Odd John, an obscure and short-lived villain who could control insects and mutate them into super-bugs.

It strikes me as a waste of potential that entomologists have never been able to turn their insights into superpowers. There are tropical termites that shoot acrid solutions out of the tops of their pointed heads, odd carnivorous creatures called berothids that live underground and capture prey by releasing toxic, paralyzing fumes, and oil beetles that shoot droplets of toxin-laden blood out from each of their leg joints. Any of these abilities would make for some impressive superhero antics. Comic books may actually turn out to be a great way to educate the general public about the remarkable abilities of insects. Hey, don't scoff—a lot of people read comic books. After all, it's not like there's an A-Plus Entomology Book and Monographs shop in town for people to flock to and buy their favorite out-of-print collectible books about insects.

Inquiring minds want to know

Given the frequency with which people cross paths with cockroaches, one would hardly think that such close encounters would be considered newsworthy. It's therefore somewhat surprising to see how frequently cockroaches turn up in tabloid newspapers. Their exploits, as recounted in the tabloids, are certainly surprising to entomologists. It's not that I'm a subscriber, or even a regular reader, of tabloids. But I do go to the grocery store and wait in long check-out lines and, like everybody else (even, I suspect, Nobel laureates), I can't help reading the headlines. When I notice an entomological headline, I confess that I slip a copy in among the cans of cat food and quarts of milk.

For example, cockroaches made the front page one day of the *Sun*. Right underneath the banner headline ("Missing baby found alive inside pumpkin") was the headline "Killer roaches invade home and attack family." The story described the terrifying experience of coffee grower Santiago Morales, whose Venezuelan home was stormed by "an army of killer cockroaches." Morales speculated that a powerful storm flooded the coffee plantation, driving the cockroaches into his house. Although Morales, his wife, and his two children survived the attack, the family dog was not so fortunate. The dog succumbed to the ravages of "dozens of greedy gobbling roaches," proving yet again that coffee and cockroaches do not mix.

Even more insidious than such natural disasters as coffee-crazed cockroaches are the ones engineered by unscrupulous insect trainers. The *Sun* also carried a story titled, "My husband trained roaches to attack me . . . claims terrified wife." In Toluca, Mexico, Roberto Guarvez "groomed a houseful of cockroaches into fighting-and-biting machines that would attack only his wife as she slept helplessly in bed." Regina had her suspicions after awakening on three separate occasions covered with cockroaches while her husband slept undisturbed, but she didn't fully comprehend her husband's involvement until she actually heard him ordering the cockroaches to kill her. Evidently, in Mexico, a person cannot be booked on charges of assault with a deadly arthropod, so Regina Morales filed for divorce on the grounds of mental cruelty.

At least one judge would have been sympathetic. "Judge attacked with roaches!" according to an edition of the *Weekly World News*. Maria Terwen, of Berkeley Springs, West Virginia, "dumped thousands of cockroaches" on the desk of Judge Margaret Gordon to "emphasize bad living conditions in her apartment complex." Maria Terwen received a contempt charge for her efforts. Although no details are provided, I would guess that the attorneys present in the courtroom allowed the cockroaches to scuttle off without smashing them, as a professional courtesy.

Elisabeth Muller's marital problems, while not in the same class as Regina Morales' woes, were nonetheless deemed newsworthy by the editors of *Weekly World News*. Her "wacky hubby eats roaches—he mixes those creepy critters with cereal, muffins and even meatloaf, says his fed-up wife." Werner Muller, a biologist from Hannover, Germany, considers cockroaches "one of nature's most perfect foods. . . . They're a superb source of protein, one of nature's best-balanced snacks." Since they're not available at local grocery stores (at least, they're not *for sale* at local grocery stores), Werner maintains his own cockroach supply by breeding them in shoeboxes in his garage. The Mullers' marriage, you will be relieved to hear, is not at risk, since Werner has promised not to eat cockroaches or talk about them in front of Elisabeth (although he did pose for *Weekly World News* cameras for three different photos, downing cockroaches on cereal, on pancakes, and right out of the shoebox).

Cuban refugee Jorge Torres of Miami, Florida, who installs fire sprinklers for a living is, like Werner Muller, favorably disposed toward cockroaches, although as a source of inspiration rather than a source of nourishment. As reported in the *Sun*, cockroaches helped him "win lottery millions." Torres used "special Cuban symbols" to pick numbers—according to his system, lucky number 48 is symbolized by cockroaches. Speaking through an interpreter, Torres claimed that he actually didn't like cockroaches but thought he'd "take a chance" on cockroach number 48. His decision was worth $6,230,000, with which Torres intended to buy a Corvette and a new house. No mention is made in the story of sharing his winnings with the local cockroach community.

Perhaps the most surprising of all cockroach stories in past issues of tabloid newspapers was the one from *Weekly World News* for January 15, 1991. A two-inch headline under a foot-long photograph of a cockroach proclaims, "Don't stomp on this guy—he's your kissin' cousin! Cockroaches and humans are kinfolks, says expert." Although the expert in the story is unnamed, he is reported to be one of the "government scientists who've spent a lot of time smashing roach heads and checking out all the bug juice" who went on to author a "startling report in the scholarly journal *Science.*" As far as I can determine, that would be either Ronald Nachman, G. Mark Holman, William Haddon, or Nicholas Ling, who published a paper entitled, "Leucosulfakinin, a sulfated insect neuropeptide with homology to gastrin and cholecystokinin" in *Science* in 1986. Nachman and company isolated the substance leucosulfakinin from extracts of three thousand *Leucophaea maderae* heads, purified it, and sequenced it, to discover substantial homology with the carboxy terminus of the human brain-gut hormones gastrin II and cholecystokinin. In fact, six of eleven of the amino acid residues of leukosulfakinin match those in gastrin II, "the highest percentage reported between insect and vertebrate neuropeptides." This striking similarity prompted the authors to suggest that the peptides "are evolutionarily related."

The *Weekly World News* reporter described these findings differently: "The next time you stumble into the kitchen for a late night ham on rye and come eyeball to eyeball with the ugliest bug in the universe, don't start squealing, swatting, and swearing. Just pucker up, pal. Scientists have discovered that you and that creepy old cockroach are kissing cousins. . . . You and those garbage-gobbling stomach churners under the cupboard are descended from a common ancestor—and you're more closely related than you'd care to admit."

While a tad on the sensational side, the report is for the most part recognizable and even accurate, right down to the fact that it does take a lot of time to smash roach heads and check out bug juice. It's gratifying, on the one hand, to see the popular press running stories based on research reported in *Science*—even 4-1/2 years after it gets published. These days, when more than half of all scientific papers go uncited, even by other scientists, mention in the popular press is real recognition. On the other hand, it's definitely unsettling to discover that tabloid stories are based on real incidents. I had always kind of dismissed the entomological accounts along with the stories of Bigfoot babies and Elvis sightings as fantastic. Now that it appears that at least some of the reports are firmly based on legitimate refereed scientific literature, I'm going to have to reevaluate my attitudes toward the tabloids. In fact, next time I buy one, I just might pick up a lottery ticket, too.

"Let me tell you 'bout the birds and the bees..."

I've never been what you could call "cool"—ever since junior high school, I have been pretty much totally unaware of changes in taste, style, and fashion. Fortunately, being cool does not appear to be a necessary prerequisite for a successful career in entomology. There are occasions, however, when being just a little cool could be a professional boon. Every spring, I teach a general education course in entomology for nonscientists. I have discovered over time that the majority of the 150 or so students who take this course each year sign up for one of three reasons:

1. the class fits their schedule
2. they think it will be easier than the general education courses in physics or chemistry
3. the class fits their schedule AND they think it will be easier than the general education courses in physics or chemistry.

The point of the course is to teach students about the biology of insects (and thus about the science of biology in general) by relating aspects of insect biology to their lives. Because general education students come from across the entire campus, this objective is differentially easy to achieve. Students from life sciences, premedical and preveterinary curricula, and the allied health professions readily accept the relevance of insects to their lives; even engineers can see the connections. It's a tougher sell, however, to students in the humanities. As a result, in addition to the usual sorts of lectures about insect behavior, physiology, and classification, the course includes lectures on cultural entomology—insects in art, music, literature, history, and the like. I have to say that, of all of these lectures, the most challenging one for me to give is the one on insects in music. The challenge arises from the fact that I am basically uncool, as far as music is concerned.

When I first began teaching this course, I based my lecture on the music I was personally most familiar with—mostly folk and pop songs of the sixties and early seventies. There was certainly no shortage of material to cover. Practically all of the sixties acts had insect-related songs in their repertoire: the Beach Boys had "Wild Honey," Herb Alpert did "Spanish Flea," the Beatles performed "A Taste of Honey," and even Elvis got into the act (albeit in 1958) with a number called "I Got Stung." And most of these songs weren't obscure in their day: Bob Lind's "Elusive Butterfly of Love" hit Number 5 on the pop charts in 1966, and Jewel Akens' "Birds and the Bees" reached Number 3 the year before. But after a few years of trying to explain to the students who Donovan was (of "First There Is a Mountain" fame, with the verse "Caterpillar sheds its skin/ to find the butterfly within"), I realized I was failing to reach them. In fact, playing Burl Ives' classic rendition of the folk song "Blue-Tailed Fly" (a.k.a. "Jimmy Crack Corn") caused so much eye-rolling and gagging that I feared momentarily the class had somehow become demonically possessed.

Fortunately, fellow entomologist Deane Bowers, from the University of Colorado, took pity on me and provided me with a temporary solution to my musical problems. She assembled a tape with an assortment of songs that were much cooler than the ones I had been using. At least, I assumed they were cooler, since I had never before heard of most of them (I actually hadn't heard of some of the artists before, for that matter). Students generally agreed, and responded enthusiastically to "Hey There Little Insect" by Jonathan Richman and the Modern Lovers ("Don't land on me baby and bite me, no"), "Animotion," by Obsession ("I will have you like a butterfly. . . . I will collect you and capture you . . ."), "Tsetse Fly" by Wall of Voodoo ("I'm feeling kind of sleepy now/I was bitten by a tsetse fly"), and "King Bee," by the Rolling Stones ("I'm a king bee, baby, buzzin' round your hive . . ."). The tape, however, was made in 1987, and, as far as the students were concerned, had become a golden oldies compilation by the time the Democrats had regained the White House in 1992.

A term paper requirement provided a more permanent solution to shifting cultural standards. All students have to turn in a term paper for the course on the subject of their choice, and I encourage them to select an aspect of insects and culture about which they know more than I do. Not surprisingly, the role of insects in popular music has turned out to be a popular topic. This is how I have come to possess over a dozen audiotapes of collections of insect songs organized around various and sundry themes. I have a particularly impressive collection of insect punk and grunge rock songs. Not only are these artists and songs I would never otherwise have heard of, these are entire genres I would never otherwise have heard of. The subtle taxonomic distinctions between "hardcore/funnypunk," "hyperoffensive

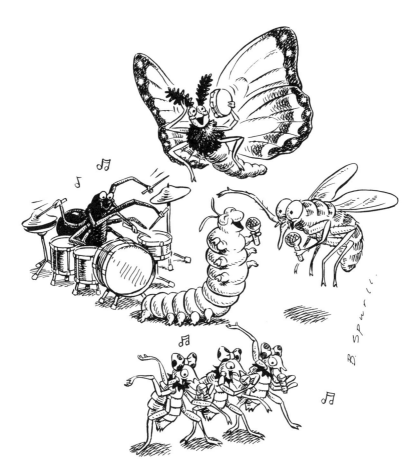

"From the top, in the key of bee-flat."

punk" and "neopsychedelic post-punk" I confess escape me entirely. Any entomologist over the age of forty will be absolutely astounded to discover the frequency with which insect references appear in contemporary music. The nineties may even qualify as an arthropod musical golden age.

Calls for celebration, however, may be a little premature; musical insect references just aren't what they used to be. Back in the sixties, insect songs were happy and pleasant, with titles like "Bumble Boogie" (B. Bumble and the Stingers, 1961), "Sugar Bee" (Cleveland Crochet and Band, 1960), "Butterfly Baby" (Bobby Rydell, 1963) and "Funny Little Butterflies" (1965—by Patty Duke, of all people). Today, insects appear in songs with titles that are for the most part unprintable here. Among the less offensive more or less contemporary entries with lyrical references to insects are "Let's Lynch the Landlord" by the Dead Kennedys ("There's rats chewing up the kitchen, Roaches up to my knees"), "Fly on the Wall" by Jesus Lizard ("That damned fly I told you about is keeping me up again"), "Bugs" by Adrenalin O.D. ("Armies of bugs, training to attack. . . . Coming out of the woodwork, coming through the floor . . ."), "I am the Fly" by Wire ("I'll shake you down to say please/As you accept the next dose of disease"), "Mindless Little Insects/Too Many Humans" by No Trend ("You breed like rats . . ."), "Moth in the Incubator," by Flaming Lips (no printable lyrics), and "Crabby Day" by Pansy Division.

"Crabby Day" is an account of an infestation of pubic lice acquired after a one-night stand:

"To sleep with him I was excited
But later I was less delighted
When I found that special present
He left in my pubic hair."

The biology in the song is most accurate but I have trouble lecturing about the song without cringing between verses and wondering when I'm going to start getting phone calls from incensed parents. Possibly the song least suitable for lecture purposes is a song from the group Ween, the title of which can be loosely rendered as "Flies landing on my uniquely male body part." As far as I can tell, it's a love song but then again I'm so uncool I might have completely missed the point.

So, times have changed and insects in music have changed with them. I know it's not cool to say so, but I'm not convinced the change has been for the better. I miss the oldies, collectively light-hearted and vermin-free as they were. While kids today may not enjoy listening to "Blue-tailed fly," I've always liked the song, and I'm deeply grateful that Burl Ives managed to sing a song about flies without dragging his private parts into it.

Bizzy, bizzy entomologists

A chance conversation with a colleague who happened to be teaching a course on women in science got me to thinking about game-playing. She showed me a passage from an article by J. B. Kahle (1985) evaluating "factors contributing to the under-representation and under-utilization of women in science." Among other things, the author attributed disparities in computer literacy between boys and girls at least in part to a male bias in available computer software; "out of seventy-five titles, appropriate for middle-school children, more than a third have been rated as being exclusively for males. Only four titles, 5 per cent, have been identified as being of primary interest to girls." On reviewing those titles identified as being of primary interest to girls, I gained little insight on how the author did these identifications; as a former middle-school girl, I can't recall ever being even vaguely interested in "Typing Fractions." Notwithstanding, it occurred to me that a similar phenomenon might account for the "under-representation and underutilization" of entomologists in science. Think about it—we all have high school classmates who went on to become doctors, but how many of us can name high school classmates who went on to become entomologists? There's probably a higher probability that we have classmates who went on to become serial killers than went on to become entomologists. I thought that maybe, at some critical juncture, young boys and girls simply weren't given toys appropriate to piquing their interest in the professional possibilities of working with small, crawling animals.

So I went to the toy store. My suspicions that toy manufacturers are not targeting the entomologically inclined were instantly confirmed. At first glance, some games may actually seem to have educational value. Take the venerable Milton Bradley game for preschoolers, Cootie. The object of the game is to be the first player to construct a Cootie Bug from various and sundry Cootie pieces (body, head, antennae, eyes, tongues, and legs). Anatomical relationships are even fairly accurate—a complete Cootie Bug has three clearly discernible body regions, three pairs of legs, one pair of appropriately located antennae, and a nicely coiled proboscis, as do so many flesh-and-blood (or cuticle-and-hemolymph) insects. So why do I feel that this game might not entice youngsters to further their entomological educations? Basically, because they're COOTIES, for crying out loud. What kinds of conversations are transpiring all over the country, probably every day?

Six-year-old: "What exactly is a Cootie Bug, Mommy?"
Mommy: "Well, dear, a Cootie Bug is a body louse—a repulsive, disease-carrying ectoparasite that lives under your clothing and sucks your blood."
Six-year-old: "Waaaaaaaaaa!"

Milton Bradley, by the way, now sells a Giant Cootie Game, so kids can assemble even larger disease-carrying ectoparasites that live under clothing and suck blood, as well as a game called Ants in the Pants, the object of which is to propel, tiddlywink-style, sixteen ants into a pair of oversized pants with suspenders, and yet another, called Bedbugs, which consists of removing jumping bedbugs from the bedclothes of a nightgown-clad sleeper. These games seem more likely to promote insomnia than they are to promote a healthy interest in insects.

Other arthropod characters popular with game designers are slavering vicious predators. Milton Bradley offers Spider Wars, "the Spider Fighter Game," the object of which is to "knock all of your opponent's spiders off the web or get to your opponent's nest first." The back of the box exhorts youngsters to "push your spider legs in and pop your opponent's spider legs out! Make enemy spiders dangle, then knock 'em off the web!" I can't help wondering whether kids who enjoy these sorts of games might well be the sort to grow up and eat their mates. TSR Incorporated offers Web of Gold, a treasure hunt through a gold mine that features "a giant spider, one of the terrifying creatures that drove away the conquistadors and inspired the legends. Spiders are not interested in gold, only in silencing the noisy intruders who disturb the quiet of their cold and lightless lair." The box cover features an irascible-looking tarantula, ill-tempered, no doubt, at being depicted dangling from an orb web inasmuch as tarantulas never spin aerial webs.

More egregious than even these spider games (which, remember, exhort players to pop the legs off arthropods) are the cockroach games. Now, admittedly, I'm no great admirer of cockroaches, but even they don't deserve what they get from Milton Bradley. Milton Bradley markets a game called Splat— "The Bug Squishing Race Game," the object of which is to "get 2 of your bugs to the midnight snack before they get splatted." To play the game, players make their own bugs (with the plastic Bug-O-Matic provided) out of "colorful Squish-It dough."   Large, hand-like objects are provided to do the squishing should a bug be unfortunate enough to land on the wrong square. Along the same lines, from Iwaya Corporation in Taiwan, is Wacko the Cockroach, a portly, furry   winged insect-like creature with the requisite number of antennae that squats engagingly on the floor with a sort of sleepy-eyed expression. Accompanying Wacko in his box is a long-handled plastic mallet. The object of the game (I kid you not) is to smash Wacko with the mallet, causing the battery-powered cockroach to make plaintive squeaking noises and to run frantically around in circles. The box explains in fractured unpunctuated haiku-like verse:

"I'm WACKO THE COCKROACH, the
Toughest roach ever you can hit me on the
Head with my own hammer I'll run away
Squealing, but I'll be back I have courage,
It's brains that I lack!"

I bought Wacko the Cockroach for my then three-year-old nephew Adam; my sister said he cried when his father, demonstrating the game, slammed the mallet down on Wacko's head. Little Adam never played with it again (although my brother-in-law ordered another one for himself).

About the least offensive role that insects play in children's games is as prey items. Such is the case for Milton Bradley's Melvin the Motorized Looney Gooney Bird, who gobbles up plastic antlike creatures, and for Mr. Mouth, The Feed the Frog

Game ("flip flies into frog's snapping mouth and his eyes jiggle and his head spins round and round"—not exactly a reaction to promote experimental entomophagy). Tyco's "Grabbin' Grasshoppers the Jumpin' Grasshopper Game" allows players to wait expectantly for spring-loaded grasshoppers to leap up off the playing surface and snag them with insect nets.

The one saving grace in this great arthropod toy desert, the one game that restores my faith in an industry that is ostensibly dedicated to educating impressionable youth, comes from Waddingtons Games Ltd. of Leeds, England. Bizzy, Bizzy BumbleBees is the "Bee-Boppin' Pollen Poppin' Race Game" (note: games manufacturers seem to drop the final 'g' a lot; viz., Grabbin' Grasshoppers the Jumpin' Grasshopper Game and, right next to it on the shelf, Superdough Sparklin' Butterfly Maker). Up to four can play. Each player equips himself with a beehive and a colored headband, attached to which is a matching magnetic bumblebee. The object of the game is to pick up steel pollen marbles from a rocking flower with the magnetic bumblebee attached to your head and place them in your beehive until all of the marbles are gone. The bumblebee that picks up more pollen marbles than anyone else is the winner.

This game is obviously extraordinary in many ways. First of all, nobody gets squashed, swallowed, or mutilated in the course of playing, although, one runs the risk of losing one's marbles. Secondly, the players' objectives are actually constructive. Along those lines, according to the instructions ("Bee-4 you play"), all players must take the "Bumblebee Oath"—

"I promise NOT to purposely hit someone else's bumblebee with mine."

"I promise NOT to slam the flower with my bumblebee."

"I promise not to laugh too much at how silly grown-ups look playing this game"—

words all of us surely can live by.

Actually, the Bumblebee Oath is probably the best preparation for a career in entomology. Lawyers, doctors, and business executives rarely are called upon to engage in professional activities that may leave them open to public ridicule—but it's really hard to keep one's dignity upon being confronted while swinging a butterfly net chasing nearly invisible insects on a blazing hot day illegally parked on a highway overpass. "Well, officer," you can always say, "sometimes grown-ups look silly playing this game."

P.C. insects

California has a well-earned reputation for cultural innovation. So it's not altogether inconsistent that, way back in 1929, it was the first state to officially designate an insect as a state symbol. Even as long ago as 1929, state flowers and trees were old news—Washington, Delaware, Oklahoma, Maine, Montana, and Nebraska had all selected state flowers even before the turn of the twentieth century. State birds were also fairly widespread by 1929. But, because of California's progressive thinking, state insects became a reality decades before there were any state dogs (Chesapeake Bay Retriever, Maryland 1964), state drinks (tomato juice, Ohio 1965), state vegetables (chile and frijole—New Mexico 1965), state shells (Scotch bonnet, North Carolina 1965), state horses (Appaloosa, Idaho 1975) or state sports (jousting, Maryland 1962). In fact, the California state insect antedates by more than 20 year the state "animal," the California grizzly bear. State animals, by the way, are all fur-covered creatures with mammary glands; in the political arena, insects, fish, birds, reptiles, amphibians, and anything extinct don't count as "animals."

The idea for a state insect in California came from the Lorquin Entomological Society of Los Angeles, whose members sought to have their home state "be the first in the entomological field to record a local symbol of its science" (Gunder 1929). They selected three butterflies to be placed on a statewide ballot, sent to "every known person really interested in entomology in the state." A whopping 88 ballots were received, revealing a landslide 77 vote victory for *Zerene eurydice*, the "California Dog Head" or "Flying Pansy," over its opponents, the California Sister, *Heterochroa californica* (11 votes) and the Lorquin's admiral, *Basilarchia lorquini* (no votes). That the Lorquin's admiral received no votes suggests that even the person who nominated the butterfly for the ballot didn't even vote for it.

Times have certainly changed with respect to insect politics, regrettably for the worse, since that landmark election. I personally can't see anything named "Dog Head" or "Flying Pansy" winning any kind of election nowadays. State insects are now no longer chosen by powerful influence groups like the Lorquin Entomological Society of Los Angeles—rather, they are generally selected by statewide balloting of schoolchildren, who are of course otherwise not allowed to vote on any issues of substance. As I write, there are 27 states with official state insects and two states with arthropods as state fossils.

Given that there are at least 30,000 species of beetles alone in the United States, it would seem statistically unlikely that any two states should end up with the same state insect. It's not at all surprising, for example, that seven states have chosen "milk" as their state drink; there are simply not that many substances humans imbibe that can be regarded as wholesome and non-addictive. Yet 12 of the 27 states with official state insects have named the "honey bee" (an introduced species) as their state symbol (13 if you count Utah, which doesn't have a state insect but which calls itself the "Beehive State"). These states—Arkansas, Georgia, Louisiana, Maine, Mississippi, Missouri, Nebraska, New Hampshire, New Jersey, South Dakota, Vermont, and Wisconsin—range geographically and politically all over the map. It's unlikely that voters in these states could agree on any issue other than state insect. Five more have named the "ladybug" as their choice—Delaware, Massachusetts, New Hampshire, Ohio, and Tennessee (although Tennessee appears to have named the firefly as its state insect as well). Distributed among the remaining states are the "firefly" (Pennsylvania and the other half of Tennessee), the European praying mantis (Connecticut), the Baltimore checkerspot (Maryland), the "dragonfly" (Michigan), the monarch butterfly (Illinois), and the Oregon swallowtail (Oregon).

As far as I can tell, California was unique in having a state insect until 1973, when the forward-thinking people of Maryland designated the most appropriate Baltimore checkerspot as state insect (but I would expect nothing less from a people who designated jousting as their state sport). The inspired choice of the people of Maryland stands in stark contrast to the actions of residents of Arkansas, Nebraska, New Jersey, and North Carolina, who within the next three years all named the honey bee as the unique and distinctive symbol of their respective states.

Unfortunately, the concept of naming an insect to represent the spirit of the state seemed to have gotten lost in the political process. The proliferation of beneficial insects (bees, mantids, and ladybugs) among the ranks of state insects demonstrates this point dramatically. By and large, state flowers are not useful; otherwise, the soybean, and not the violet, would be the state flower of Illinois. By and large, state insects are useful. Why is it that only useful insects are "politically correct"? Admittedly, it is not only insects that find the political environment hazardous. When New York was preparing to name the beaver their state animal in 1975, a disgruntled legislator in Oregon (which had designated the beaver as its state animal six years earlier) suggested that "the cockroach" might be a more fitting symbol for the Empire State. And in 1987, Governor Jim Thompson vetoed a bill that would have named the Tully monster *Tullimonstrum gregarium* the state fossil of Illinois because "an election among schoolchildren . . . would resemble the Soviet electoral system." The governor was bothered by the proposed election process, according to which schoolchildren would receive a ballot to mark either "yes" or "no" to the Tully monster, a 300 million year old marine animal known only from coal shale deposits in Illinois. The *Champaign-Urbana News Gazette* (September 16, 1987) quoted the governor as eloquently and persuasively arguing that "It's either yes or no on the Tully monster. That's un-American. . . . That's how they run elections in Russia. This is not Russia, it's Illinois." Interestingly, the Tully monster finally did win approval as state fossil in December 1989 and Thompson declined to run for re-election three years later. It remains to be seen whether Governor Thompson will be nominated for anything in 300 million years.

That useful insects are overwhelmingly the choice of most Americans to represent their place of residence is not for want of trying by more entomologically enlightened voters. In the discussion over the Wisconsin state insect in 1985, for example, the mosquito was unsuccessfully pitched as a possible candidate due to its status as an important link in the aquatic food chain. During the long debate over a state insect for New York, legislators were unconvinced by conservationists who argued that an endangered species, such as the Karner blue butterfly *Lycaeides melissa samuelis*, would be a more appropriate symbol of the state's natural resources than the ravening, aphid-devouring ladybeetles. In 1992, it was a politician who showed some imagination facing down the "bee boosters" in the Oklahoma Legislature when he proposed "the tick" as state insect. Although his taxonomic skills weren't exactly awe-inspiring, inasmuch as ticks are arachnids and not insects, his logic was impeccable.   State Senator Lewis Long "lobbied for the tick because it would have something in common with mistletoe, the official state flower. Both are parasites" (*Chicago Tribune*, April 22, 1992). Senator Gilmer Capps, tool of the state beekeepers and proponent of the original bill, prevailed, however, and the honey bee bill was adopted unanimously. That unanimous ballot meant that, when the votes were counted, Senator Long didn't even have the backbone to support his own candidate.

It's true that insects have never really enjoyed good public relations—but politicians these days are hardly in a position to throw stones. What with influence peddling, pork barreling, sexual harassment, illegal campaign contributions, and all, it may well be that the next nominee for state insect will be a spineless politician.

Over-the-counter insects

I'm not really much of a sports fan, and women's track and field events in particular have never held any great fascination for me—no doubt the legacy of a sadistic tenth grade gym teacher, who used to send us back to the building from the school track with the promise to give C's to the last four people through the door. My interest was piqued, however, in 1993, when a controversy erupted during China's seventh National Games. In case you were following the pennant races at the time, I'll recapitulate—on Saturday, September 11, Qu Xunxia shaved more than two seconds off the world's record for the women's 1500 meter race, only three days after teammate Wang Junxia broke the record for the women's 10,000 meter race by 42 seconds. Critics quickly accused the team of using performance-enhancing drugs. According to one newspaper account, the team's coach, Ma Zunren, rose to his team's defense at a news conference: "he held up a light brown box and said, to laughter and the excited clicking of cameras, that the key to their success is a health tonic made from caterpillar fungus."

This kind of sports news I can relate to. I've actually had first-hand experience with Chinese medicinal insects—at least buying them, if not exactly trying them. My first brush with Chinese insect materia medica was during the Tenth International Congress of Entomology in Vancouver, British Columbia, which also happened to be my honeymoon. Okay, so maybe giving two talks at an International Congress of Entomology is not everybody's idea of a romantic honeymoon venue, but it seemed like a good idea at the time. One day between sessions, my husband Richard and I wandered over to Vancouver's Chinatown and stumbled across a Chinese herb shop. Looking in the window, I saw what I recognized as a jar full of cicada exoskeletons prominently displayed; we went in, and, with a lot of gesturing and pointing, conveyed our interest to the non-English-speaking proprietor and headed off triumphantly with our purchase.

The next shopping expedition, in Honolulu a year later, met with substantially less success. In the midst of a Pacific Rim Chemical Congress, we headed for Chinatown and quickly found a Chinese herb shop. This time, however, there were no cicadas in the window and, instead of merely pointing, we were forced to explain to the non-English-speaking proprietor that we were looking for insects. After numerous gestures, flitting motions, and buzzing sounds, the proprietor, comprehending what we were looking for, stared at us as if we were nuts (this from a guy with dried lizards in his window), and motioned for us to try the shop across the street. Whether he phoned ahead to warn his competitor, we'll never know, but we had even less success at the next shop. This time, we got nothing but blank stares until I took out a piece of paper and pencil and drew a crude picture of an insect. The proprietor looked at us in horror, shouted something that I presume was unflattering in Chinese, and hustled us out the door. Standing out in the street, Richard thought for a minute and then figured that the proprietor probably assumed we were from the Health Department there to accuse him of harboring cockroaches. Needless to say, we left Honolulu bereft of cicada exoskeletons or caterpillar fungus, although Richard did manage to pick up some nice eelskin wallets for next to nothing in the gift shop next to the hotel.

I thought my insect materia medica buying days were pretty much at a standstill until I went to the mailbox one day and found a catalogue from an outfit called Standard Homeopathic in Los Angeles, (where else) California. I hadn't ordered the catalogue, but I imagine I received it as a result of being on a number of very strange and not necessarily complementary mailing lists. I wonder sometimes what the mailman thinks when he delivers the *Vegetarian Times* and the *Omaha Steaks International* catalogue to the same person. Homeopathic medicine, I read from the enclosed brochure, is a "therapeutic system" developed during the early nineteenth century by Samuel Hahnemann in Germany. It's based on the ancient dogma, *similia smilibus [sic] curentur*—"let likes be cured by likes." In a nutshell, the idea is that substances that in large quantities cause illness in humans can, when administered in much smaller doses, effect a cure of that same illness. "In a nutshell," indeed, since most of the medicaments in the catalogue were of plant origin—a case in point being the use of *Nux vomica*, the "poison nut," for "gastric and living disorders occasioned by high living, overeating, or excessive medication. . . . *Nux Vomica* can also help break the laxative habit." Scattered in amongst the plants, however, were some familiar names—among them, the honey bee *Apis mellifica* (an almost familiar name at least—entomologists spell it *Apis mellifera*), recommended for edema, insect bites, and skin problems, the Oriental cockroach *Blatta orientalis,* to be taken for cough, the Spanishfly *Cantharis*, recommended for burns and sore throats, and the hornet *Vespa crabro*, prescribed for nausea and burning skin eruption. I was ecstatic—here was an opportunity to order insect-derived medicines in the comfort of my own living room, without having to undergo the fuss and bother of traveling to inconveniently far-off places like Hawaii.

Ordering these products was about as easy as communicating with Chinese-speaking proprietors of herb shops, however. These remedies were available either as tinctures, tablets, or pellets, in various potencies. *Apis mellifica*, for example, was available in all three forms. The remedies were also available in a variety of formulations, ranging in potency from 3X (a 1 to 1,000 dilution) to 30X (a 1 to 1,000,000,000,000,000,000,000,000,000,000 dilution). I just kind of closed my eyes, filled in the form, and hoped for the best. Several weeks and about $80 later, I received my bug drugs, no doubt raising the eyebrows of our mailman once more.

Most of the order arrived as promised, although, to my disappointment, *Cimex* (a medicinal preparation featuring bed bugs) wasn't available and *Cantharis* tincture was available by "prescription only." I still haven't used any of these homeopathic medicines and I probably won't. I probably won't even order them again. If I feel the need for homeopathic entomological medicaments, I've actually succeeded in locating a local source for some of these products—a neighborhood grocery store with a health food section stocks a wide assortment of homeopathic remedies, including a product called Flea Relief, from Dr. Goodpet Laboratories, Inglewood, where else? California, which contains, among other things, *Apis mellifica* 3X and *Pulex irritans* (the human flea) 12X.

I was disappointed, though, not to receive the *Cantharis*. After all, this is the one insect medicament with demonstrable pharmacological activity. While entomologists know it as *Lytta vesicatoria*, in popular parlance, *Cantharis* is known as Spanishfly. It's probably the world's best known and most widely abused aphrodisiac. As do many meloids, known collectively as oil beetles due to their propensity for exuding toxic oily body secretions through their joints, *L. vesicatoria* produces powerful defensive secretions, in this case containing the terpene anhydride cantharidin. Cantharidin is a potent vesicant, or blistering agent, and irritant of mucus membranes. Thus, cantharidin has dramatic physiological effects upon ingestion, which, depending on one's personal proclivities, can be regarded as either desirably stimulating or undesirably painful. Cantharidin's use as an aphrodisiac is fundamentally unsound, however, because it's extremely toxic even at low dosages—as little as 30 mg can prove lethal (and indeed, the infamous Marquis de Sade was prosecuted in 1772 for poisoning several prostitutes by administering Spanishfly to them without their knowledge). Needless to say, the use of Spanishfly and cantharidin for treating humans for erectile dysfunction has been illegal since the nineteenth century, although it is still routinely prescribed for warts.

I may have an opportunity after all, however, to get my hands on some Spanishfly. The mailman came by and left a catalogue for my husband from an outfit called Leisure Time Products (located, surprisingly, not in California but right next door in Gary, Indiana). I won't tell you what's on the cover, or what's on most of the pages, although I will say the word "hot" shows up on an inordinate number of pages. I did ask my husband how he happened to receive the catalogue, though, and he shrugged, looked innocent and said, "Mailing lists—*you* know. . . ." I will say, though, on page 47, there was an advertisement for a product called "Spanish Fly." I can't help but wonder what it really is, since the over-the-counter sale of Spanishfly for human use is strictly forbidden by law (but, I guess, so are a lot of other things depicted in that catalogue . . .). I'd really like to order it, just out of scientific curiosity, but I'm hesitant.   For one thing, I can't imagine what our mailman will think if I actually order something from Leisure Time Products.   Second, it's $12.95 a bottle and I'm not sure I want to spend that kind of a money, particularly if it's lining the pockets of a less than savory operation.   More important, though, I don't know how I'll ever decide which one of the 15 flavors to order.

Roach clips and other short subjects

Unlike many Americans, I do not view my computer as a recreational device. Actually, I tend to view it as more of an electronic tyrant that constantly makes unreasonable demands of me and that will keep tormenting me until I no longer have the strength to lift my fingers to the keyboard. While the amount of paper that can stack up on my desk is finite, as determined by such physical laws as gravity and friction, my computer seems to have a limitless capacity for storing things that aren't yet finished. Even when it's turned off, screen black and empty, it is a silent reminder of my inability to fulfill my obligations. I expect I owe an e-mail reply, a letter, a manuscript review, or a book chapter to at least 30% of the people reading this book.

So it goes without saying that I don't cruise the information highway casually. When I do venture out onto the Web, it's almost invariably for some work-related purpose. There's certainly no shortage of entomological information on the Web. If you conduct a search on the subject of "insects" using almost any of the popular search engines, you end up with a list of some 300,000 items to sort through. Here's another confession—I am not at all proficient at dealing with this information in byte form rather than book form. Books are nice and predictable; if you found one in the library five years ago, odds are good that today, barring vandalism and budget cuts, it's still sitting approximately where it was and it still looks pretty much how it looked five years ago. On the Web, nothing stays put for very long. Finding information on the Web is a lot like looking for cockroaches in an urban apartment—you know they're there, and it takes a while to find them, but, when you do, you eventually find more of them than you ever imagined possible.

I guess it's not surprising, then, that cockroaches and the Internet are so sympatico. Querying the search engine Alta Vista about "cockroaches" yields some 30,000 matching items. Conspicuous among these is Cockroach World, the self-proclaimed "Yuckiest site on the internet" with all manner of information, including video and sound files for the stouthearted. The site "How to care for pet cockroaches" offers tips on housing and feeding pet cockroaches at home. You can find out why, unless you're really fond of cockroaches, you might want to avoid staying in the Jolly Swagman youth hostel in Sydney, Australia; you can go about designing your own integrated cockroach management program; you can obtain expert opinions on cockroach allergies from the National Jewish Center for Immunology and Respiratory Medicine; and you can even look up every reference to cockroaches in the collected work of Monty Python, if you're so inclined.

But the home page that forced me out into the Internet in the first place was "Joe's Apartment." "Joe's Apartment" is a home page devoted to a feature film of the same name—the first feature film to be produced by MTV, the cable station best known for rock videos, for Beavis and Butthead, and for contracting the attention span of American 18-year-olds to about three minutes. "Joe's Apartment," an expanded version of a three-minute short by the same name first seen on television in 1992, is basically the story of a guy from Iowa who comes to New York City and finds that the only friends he can make in the big city are the cockroaches that infest his apartment. These aren't your average New York City cockroaches—they not only speak (rudely, on occasion, as you might expect of a New York cockroach), they also can sing, break-dance, perform synchronized swimming routines, and otherwise perform remarkable six-legged feats. Although 5,000 live cockroaches, wrangled by Ray Mendez of the American Museum of Natural History, were used in the filming, many of the more complex scenes involved an impressive blend of puppetry and computer animation.

I actually didn't learn about the *making* of "Joe's Apartment" from the Website, which is mostly video clips from the movie—I learned about it from television, where I've learned so many other useful things in life (like, for example, all of the words to the theme song of "Mister Ed"). To promote the movie, MTV ran a couple of half-hour programs, one called "The Making of Joe's Apartment" and the other "MTV Unbugged," which featured clips from the films, interviews with the stars and special effects people, and clips from other great moments in cockroach cinema. It was the latter aspect of the program that especially aroused my interest, having had a longstanding interest in insects in movies. The program just didn't really do justice to celluloid roaches or even to the intellectual predecessors of "Joe's Apartment."

In fact, the first full-length feature film about cockroaches was also an animated musical. In a deservedly forgettable moment of Hollywood history, "shinbone alley," an animated jazz/opera about a cockroach, was released in 1971. Based on a book of poetry written by Don Marquis in 1927, it's the story of a poet who drowns in a river and finds his soul transmigrated into a cockroach—archy—who breaks into the office of a newspaper reporter and types poetry at night on the typewriter. His poetry and his name both are rendered in lower case because, as a cockroach in pre-computer days, he couldn't hit both the letter key and shift-lock key at the same time. The movie recounts his adventures, his friendship with an alley cat, and his philosophical musings from an insect perspective. Not surprisingly, it wasn't well received by critics—Vincent Canby of the *New York Times* called it "rather ordinary." The public wasn't exactly enthralled, either; when we showed it here at the University of Illinois at an insect fear film festival a few years back, a disgruntled audience began to chant "Die, archy, die!" before we even got to the third reel.

Things have obviously changed a lot in the last 25 years. For one thing, cockroaches are much more conspicuous in films than they've ever been. In fact, it's a little disquieting just exactly how rapidly their numbers are growing. From 1971 to 1980, there were really only two theatrical releases featuring cockroaches in pivotal roles: "Bug" (1975), about oversized flesh-eating combustible cockroaches unleashed from the bowels of the earth after an earthquake, and "Damnation Alley" (1977), about a post-apocalyptic world dominated by oversized, flesh-eating, noncombustible cockroaches. In the early 1980s, pickings were slim, the principal entry being "Creepshow" (1982), an anthology film one-third of which featured E. G. Marshall as a wealthy New Yorker beset (and obsessed) with cockroaches. Things picked up in the late eighties, with "The Nest" (1987), "Twilight of the Cockroaches" (1987),

"Nightmare on Elm Street IV" (1988), "Blue Monkey" (1988), "Deep Space" (1989), and "Meet the Applegates" (1989), all appearing as feature films with cockroaches, or their Hollywood approximations, figuring prominently in the plot. Then, in the nineties, things began to get out of hand. Along with feature films ("Pacific Heights" [1990], "Naked Lunch" [1991], "Joe's Apartment" [1996]) and animated or live-action shorts ("Juke Bar" [1990], "Joe's Apartment" [1992]), cockroaches began showing up with astonishing frequency in music videos—viz., EMF's cockroach-laden "Lies" (1991), Juliana Hatfield's "I See You" (1992), Matthew Sweet's "Time Capsule" (1992), Soundgarden's "Black Hole Sun" (1994), and Nine-Inch-Nails' "Closer" (1994).

As an entomologist, this escalating pattern looks disturbingly familiar to me. Students of introductory ecology recognize an exponential growth curve as characterizing the pattern of population growth of organisms that are not constrained by environmental limits. In fact, if the number of cockroach–related films is plotted against time in five-year increments, the resulting relationship can be described by a quadratic regression with the equation, $y = 1981 - 1.58x + 0.41x^2$. This curve can be used to extrapolate into the future, in order to estimate the number of cockroach films at a future date, given that current growth trends continue. If so, according to these calculations, by the turn of the twenty-second century (2101), we can expect to see more than 300 cockroach films coming out every five years. That works out to about 60 cockroach films a year, or five cockroach films a month, or a little more than a cockroach film every week.

With that many movies being made about cockroaches, I can't imagine there will be time for making any other kind of movie. Hollywood may well be given over entirely to the care and feeding of cockroaches to meet the demands of producers and directors, not to mention the cockroach-hungry public. There likely will be all-cockroach cable channels for those productions released direct to video. On the positive side, such a situation might mean more employment opportunities for entomologists, but overall even I think it's a dim view of the future. Although it has often been said that cockroaches will someday take over the world, I always thought it would somehow involve a nuclear blast and lethal levels of atomic radiation, not Dolby sound and Technicolor.

Got my mojo workin' (badly)

I've never really been one for playing games. I'd like to think that this lack of enthusiasm results from an overdeveloped work ethic but, in reality, it is probably a consequence of the fact that, for the duration of an otherwise completely happy childhood, I was never once able to beat my older brother Alan at a game of Monopoly, or any other board game, for that matter. However, after at least a dozen people (including my aforementioned brother) asked me if I had seen the new computer game "Bad Mojo," I felt a professional obligation to investigate, childhood traumas notwithstanding.

I wasn't a complete stranger to computer-based insect games at the time. In January 1993, while visiting my sister and her family in Connecticut, I watched in awe as Adam Escalante, my then six-year-old nephew, adroitly maneuvered his way through a game called "Battle Bugs," basically an electronic war game but with hexapod, or occasional octopod, participants who face each other on such battlefields as kitchen counters or picnic tables engaged in campaigns given colorful names like "Dessert Storm." Adam was surprised that, as an entomologist, I'd never even heard of the game and was disappointed that a Ph.D. in entomology didn't really equip me to provide him with helpful tips on how to win. Needless to say, Aunt May's stock went down a few points that day in the Escalante household. But even the shame of having disillusioned my nephew wasn't enough to motivate me to try playing the game myself. Old aversions are very hard to overcome.

There are times, however, when personal preferences must be set aside for the good of the profession and eventually I went to a local CD-ROM discount outlet to see what I could find in the way of insect-related computer games. In comparison with my earlier foray into the world of insect-related games ("Bizzy, Bizzy Entomologists"), a few contrasts immediately came to light. For one thing, computer-based games about insects are a whole lot more expensive to buy than are traditional board or boxed games about insects. In less than an hour, I managed to run up a bill of more than $182 (and this is a discount outlet, remember) for less than one full grocery-store sized plastic bag full of games. These games also cost a whole lot more to play than your basic board games. To play a game of "Cootie," by Milton Bradley, for example (list price $6.49), you supply a table or a floor; everything else you need comes in the box. To play "Elroy Goes Bugzerk," from Headbone Interactive, you need a 33 MHz processor, an 8-bit color monitor, and a double speed CD-ROM drive.

Which brings up yet another problem: compatibility. To play Cootie, it doesn't much matter if your table is made of oak or maple, or if your carpet is yellow or beige. After I spent $30 on "Bug" (Sega) and brought it home, I discovered that it was formatted for PC only—which meant that, if I were really intent on playing it, I needed to spend an additional $2200 to purchase a 486 DX4 100 MHz or a Pentium 60 MHz machine with 1 MB video RAM, Soundblaster 16, and Windows 95. Fortunately, because I possessed a System 7.1 Macintosh with 8 MB RAM, an 8-bit color monitor, and a CD-ROM drive with a 300 KB per second transfer rate at home, I was equipped, at least technologically, to play "Bad Mojo" (list price $54.95, discount price $39.95), which was, indeed, one of the games for sale at United CD-Rom. After a few false starts, which involved crashing my computer twice before I located the set of instructions for installing the software, I attempted to play the game.

At this point, the contrast between computer games and traditional games became even more stark. I really did try to enjoy it, honestly, but everything I've always hated about games that come in boxes with dice and plastic playing pieces seems to be much worse in a computer game. And there are a few new things to hate, to boot (or, I suppose, to boot up). For example, I am exceptionally prone to motion sickness and I discovered that watching images flit across a computer screen is about as effective as a commuter flight from Champaign to Chicago on a prop-jet for inducing nausea, which, needless to say, profoundly tempered my enthusiasm for playing.

I was also totally unprepared for the complexity. To find out how to play "Cootie," you read the instructions (printed in reasonably big letters) on the side of the box, with headings like "Object of the Game," "Playing Cootie" and "To Win." Pretty much everything you need to know is there, and there's really no pressing need for a toll-free help line. With "Bad Mojo," you get a few vague hints in a 14-page illustrated booklet and then you're on your own unless you want to invest in the "Bad Mojo Official Player's Guide" ($19.99), including "all the tips, tricks, and strategies you need to master this hot new game!"

I guess I should explain the concept of "Bad Mojo." Essentially, it is a role-playing game based on the character of Dr. Roger Samms, an entomologist (from the College of Chemical Ecology at the California University at Barbary Coast) with experience in pesticide development. As the game begins, he is packing his bags with large wads of cash, in apparent haste, in preparation for leaving for Mexico. After gazing into an antique locket, however, Dr. Samms is, with apologies to Franz Kafka, miraculously transformed into a cockroach. The object of the game is to help Dr. Samms make his way back to his own humanity, by directing him through a decrepit old building along San Francisco's waterfront and avoiding the thousand natural shocks cockroach flesh is heir to. These shocks include roach motels, called "cockroach corrals" in the game, voracious spiders, an overly playful cat, flypaper, vacuum cleaners, cigarette butts, cans of insecticide, rats, mousetraps, and a range of other such urban delights. It's all very mystical and mysterious; intermittently, floating disembodied heads in liquor bottles appear on screen to provide the player with poetic but cryptic hints as to how to proceed. I must confess, I gave up before my first liquor bottle (on-screen); I just found it all too frustrating.

Perhaps most frustrating was the fact that elements of the game didn't make sense, entomologically speaking. It's true that the cockroach is rendered anatomically with tremendous accuracy, even down to the level of tarsal segments, and its movements are amazingly realistic—realistic enough to fool Leo, our least intellectually gifted cat, who made playing the game considerably more complicated by periodically launching himself at the screen in an attempt to capture and eat the cockroach image. After all, the game boasts more than 800 two- and three-dimensional scenes and 35 minutes of live-action video. And a list of acknowledgments in the game package indicates that at least two entomologists were consulted in the development of the game. All the same, I found scant evidence that the entomological consultants had had much impact on the game. The on-screen cockroach simply did not behave in ways that I, as an entomologist, had long believed that cockroaches should behave. To cite just one frustrating example, according to the rules of "Bad Mojo," cockroaches cannot cross bodies of water. Had my cockroach ever reached a body of water to cross, I never would have figured out this rule—faced with a cockroach that for hours refused my command to move forward, I probably would have thrown a desk chair at the computer while citing literature that proves without doubt that cockroaches can, in fact, swim.

Fortunately for me and my computer, before my CD-ROM drive transfer rate dropped precipitously from 300 KB per second to zero, my husband heard the disgruntled mumblings and stopped in to see what my problem was. Now, Richard is quite a game player and, I feel compelled to point out, an only child without a Monopoly-hustling older brother, and he kindly volunteered to take over the cockroach for me. He spent about a week, in spare hours, manipulating the cockroach around cigarette butts and broken toilets and, eventually, managed to restore Dr. Samms to human form, long after restoring me to my human form. From time to time he'd call me into the room to show me the various situations he'd maneuvered in and out of, whereupon I'd call some petty biological detail into question; he would then shrug and continue playing and I'd stalk out of the room again, vaguely perturbed but decidedly relieved that I wasn't the one playing the game.

So ultimately it was Richard, a Ph.D. in linguistics, who figured out the game. Once again, my Ph.D. in entomology failed to provide any special advantage in tackling a game about insects. I asked Richard if he felt that he'd gained any insights into cockroach life, or even entomologists, as a result of his effort, and his answer was noncommittal (although he did seem to take special notice of an obnoxious and domineering female department head that I'm assuming is entirely unrelated to the fact that he's married to a female department head). I really am grateful to him, though, and I feel that I ought to express my appreciation in some tangible way. Maybe I can get him a Bad Mojo T-shirt ($18.95), a Bad Mojo baseball cap ($22.95), or a Bad Mojo poster ($17.95). Even better, I can get him a Bad Mojo limited edition designer watch ($59.00), so we'll be able to get an accurate estimate of just how long it's been that we've been having so much fun.

Weird Al-eyrodidae?
Weird Al-eocharinae?

It was while I was reading an obituary several years ago that I realized just how interesting entomologists can be. Dr. Robert Traub, who died December 21, 1996, at Bethesda Naval Medical Center in Maryland, was a world-renowned authority on the systematics of fleas. His obituary, appearing in the *New York Times* on January 5, 1997, recounted in detail his many professional accomplishments. Although I was reasonably familiar with his work, I was completely unaware that Dr. Traub had over the course of his life acquired some avocational interests along with his professional ones. Among other things, he was evidently an avid collector of blowguns—an avocation, I would venture to guess, unique among entomologists.

This revelation fell hard on the heels of another along the same lines. While in my hotel room in Louisville, Kentucky, reading the program for the 1996 national meeting of the Entomological Society of America, I discovered that Dr. James Slater, winner of the 1996 Founder's Award, is, in addition to being a distinguished Hemiptera systematist, a nationally renowned authority on antique glass and, in fact, is a past president and honorary member of the National Milk Glass Collectors Society. Polite inquiries around my own department reveal that just about everybody has achieved avocational accolades; among my colleagues here are a navy-blue belt in tae kwan do, a competition-class wind surfer, and a cantor at a local synagogue. And it's not just at the University of Illinois that entomologists distinguish themselves at their avocations. At Cornell University, where I was a graduate student, my major professor, Paul Feeny, was an avid sailor and one-time captain of the Cayuga Lake Cruising Fleet; another member of my thesis committee was a superb pianist and conductor of a small orchestra, and at least one additional member of the entomology faculty was a breeder of prize-winning guinea pigs.

I can't claim to have ever achieved avocational distinction, at least in part because my vocation keeps getting in the way. Because I am chronically unable to keep up with my professional obligations, I have a hard time justifying time off from work and, as a consequence, virtually all of my avocational activities somehow get connected to my profession. I collect stamps, for example, but I only collect stamps with insects on them. Moreover, I organize these stamps taxonomically, rather than by country or date of issue, which means my collection renders most real stamp collectors apoplectic with disapproval. I can't claim to be a definitive authority on films, other than those featuring oversized radiation-mutated arthropods; since, last I checked, François Truffaut and Jean-Luc Godard never made any big bug films, my opinions on film are rarely solicited by true cineastes. I'm not even an accomplished consumer; when I go shopping, I buy the paper towels with butterflies printed on them, irrespective of whether they're cheaper, stronger, or quicker picker-uppers.

I do have one legitimate outside interest, though. For reasons I'm not certain even I can articulate, I am inordinately fond of the work of "Weird Al" Yankovic. For those with loftier musical tastes, I can explain that "Weird Al" Yankovic is widely acknowledged as the nation's premier pop music parodist. Over and above his phenomenal mastery of the minutiae of popular culture and his amazing facility with language, he has a wonderfully twisted sense of humor that I find very appealing. I've known about "Weird Al" for years, but I became particularly enthralled when I discovered that Hannah, my then six-year-old daughter, enjoyed listening to his songs on tape in the car to and from school, albeit on an entirely different level, inasmuch as she tends to prefer the rapid patter and funny accents to the parody and pastiche. Thus, "Weird Al" spared me from the risk of driving into a tree to escape from endless repetitions of her previous choice of listening material, the Chipmunks' version of "Uptown Girl."

This is not to say that "Weird Al" Yankovic and insects are entirely unrelated. There are, by my count, at least a dozen references to insects in "Weird Al"'s oeuvre. There are passing references to bed bugs in "Dare To Be Stupid," potato bugs in "Addicted to Spuds," flies in "That Boy Could Dance" and in "Good Old Days," mosquitoes in "Jurassic Park," "some kind of bug" in "Slime Creatures from Outer Space," termites in "The Home Improvement Song (I'll Repair for You)" and to the Boll Weevil Monument (which really exists in Enterprise, Alabama) and tarantula ranches in "The Biggest Ball of Twine in Minnesota." In "One of Those Days," "Weird Al" recounts the terrible things that happen to him during one particularly excruciating day; these include being followed by "a big swarm of locusts" and being tied to a tree by Nazis and covered with ants. And, in a parody of Camille Saint Saens' "Carnival of the Animals," he even wrote a poem about cockroaches:

"Some think the Cockroach is a pest
But that's the insect I like best
I love the way they run in fright
When I turn on the kitchen light
And when I squish them on the ground
They make a pleasant crunching sound."

So it's not as if "Weird Al" Yankovic would likely top any entomologist's list of insect-friendly musicians. As avocations go, I had picked one that allowed for a decent separation of work and play.

In the summer of 1996, though, despite my best intentions, my vocational and avocational interests intersected. I managed to convince Richard, my longsuffering spouse, that, for my 43rd birthday, it would be a splendid idea to drive 3-1/2 hours to Rockford, Illinois, to attend a "Weird Al" Yankovic concert. When I found out that the man himself would be making a personal appearance before the concert at a local music store only a few miles from the hotel where we were staying, I also convinced Richard that it would be a rare treat to wait an hour and a half in line to meet "Weird Al." Actually, to be completely accurate, I didn't convince him, exactly; after one glance at the line, he and Hannah took off for parts unknown while I held their places in line for the hour-and-a-half. During the wait, I had time to look over the crowd and note that Rockford's entomologists had not turned out in droves for autographs. Most of the crowd actually seemed to consist of preadolescent boys accompanied by a bemused parent. I was undeterred, however, and patiently waited my turn, "Bad Hair Day" cassette tape (his latest) in hand, ready for signing.

How the world sees insects

Before I knew it, I was at the head of the line. Actually, it wasn't quite "before I knew it"—it was, I think, the longest I had stood waiting in a line, for any reason, in my entire adult life. By this time, my spouse and child had rejoined me and we all three faced "Weird Al" Yankovic in person. After he obligingly signed and returned the "Bad Hair Day" cassette, I found myself handing him a copy of my book, *Bugs in the System*. I brought it with me with the intention of offering it to him as a gift, explaining that I sincerely hoped he might get as much pleasure from my vocation as I did from his. I had rehearsed a little speech to that effect for much of the 90-minute wait. Much to my amazement, though, the speech I had so carefully rehearsed somehow degenerated into a few mumbled and largely incoherent phrases upon delivery.

To be honest, I really can't remember too many details about the exchange. I do recall that "Weird Al" looked a little surprised at being handed a book about insects—at least I think that I recall that he looked surprised. I can't imagine what he thought. For that matter, moments after the book left my hand, I couldn't imagine what *I* had thought. It's not as if rock stars are renowned for reading books about insects, particularly in the middle of a 90-city tour. After all, even college students whose grades depend on it often don't read books about insects. And it's not as if anything he had ever said or done indicated he had even the slightest interest in insects. What on earth had I been thinking of? What incredible lapse in judgment had just taken place?

That evening, we attended the concert—only the third concert I had ever attended in my life up to that point and the first since 1984, when I was somehow talked into seeing Slim Whitman perform at the Moultrie-Douglas County Fair—and we headed for home the next day. I left Rockford convinced that, by my

actions, I had reinforced every stereotype of the clueless entomologist.   I was fairly confident that the book had ended up left
behind on some hotel room dresser, if, in fact, it even made it out
of the music store, and I devoutly wished that I could have for
once in my life left insects out of the issue. Once home, I slipped
comfortably back into being profoundly unidimensional. But
hobbies are hard to shake; in between entomological pursuits, I
continued to listen to "Weird Al" Yankovic.

And, at the December 1996 national meeting of the Entomological Society in Louisville, my avocation managed to get the
upper hand again.   During the evening session of the last day of
the meeting, I ducked out early to run back to my hotel room.
That night, the Disney Channel was scheduled to show "'Weird Al'
Yankovic: (There's No) Going Home," a brand-new hour-long
special based on his summer tour. Alone in the hotel room, while
my colleagues discussed entomological matters of import,   I
watched "Weird Al" Yankovic on television, in a program that
featured a combination of concert footage and interview segments.

I felt guilty, to be sure, but not guilty enough to turn it off and
go back to the meeting. And I'm glad I didn't—I wouldn't be
telling you this story if I had.   In an interview segment ostensibly
filmed on the road in his tour bus, "Weird Al" explained what life
on tour was like.

"Nothing makes you long for the comfort of home and family
more than traveling across the country in a bus with a big smelly
rock band," he confided to his off-screen interviewer. "Sometimes,
life on the road can get a little tiresome and one way that we like
to break up the monotony is by stopping at various points of
interest along the way to pick up a few souvenirs. For instance,

"This is your brain on bugs. . ."

There are, on occasion, some benefits to waking up early on Sunday mornings to watch television with a child. Had I not been watching Sunday morning children's television programming one morning in 1998, I probably wouldn't have seen the latest public service announcement from the Partnership for a Drug-Free America (PDFA). This very worthwhile organization is dedicated to reducing drug abuse, particularly among children and teenagers, by disseminating information about the dangers of drug use, on television as well as through other media. Actually, I didn't exactly see the announcement in its entirety that morning. I couldn't watch the entire commercial because my daughter Hannah, who was in possession of the remote control at the time, decided she wasn't in the mood for public service announcements and, over my strident protests, changed the channel in mid-spot. What I did manage to see went by rather quickly, and I saw enough to realize that the spot featured a large, purple animated insect and a tag line to the effect of "I'd rather eat a big ol' bug than ever take a stupid drug."

Having an abiding interest in insect images in popular culture, I thus felt compelled to contact the PDFA directly and this organization was kind enough to send me not only the entire text of the aforementioned ad but also a full-color video along with a lot of other information. Evidently, this ad was part of a new campaign, undertaken with the help of a 195 million dollar grant from the federal government, to run anti-drug advertising in the paid media for the first time in the history of the PDFA. Included in this effort were new and innovative advertisements on television. The spot I'd seen, titled "Big Ol' Bug," features an animated little boy and an animated insect and the text runs as follows:

(MUSIC UP)
SONG: "I'd rather eat a big ol' bug/than ever take a stupid drug
Drugs aren't cool, they can mess you up at school
Drugs are a pain, they can hurt your body and your brain
A big ol' bug with an ugly mug/is better than any stupid drug
They make you sad and your parents mad
Drugs are dumb, they make you clumsy slow and numb
I'd rather eat a big ol' bug/BUG: Don't do drugs!
REFRAIN: ...Than ever take a stupid drug!"

The bug, typical of most animated insects, was unrecognizable as to taxon and was vaguely reminiscent of Tex Avery's classic cockroach characters in Raid commercials of the sixties (complete with eyes with pupils and teeth in need of bridgework). Now, I'm totally supportive of the basic message of the PDFA ad. I'm so averse to the use of mind-altering substances that I don't even consume products containing alcohol; beer, wine, and hard liquor have always tasted like laboratory solvent to me, anyway. But I found this spot a little unsettling, for several reasons. For one thing, the little boy in the ad appeared to be willing to eat his bug sandwich with the crusts remaining on the slices of bread—a prospect that is probably more horrible to some children (like my daughter) than the notion of eating an insect per se.

More importantly, I was a little confused by the particular message presented by this spot. I take it that eating a bug is supposed to be a terrible thing, pleasant to contemplate only in comparison to an experience even more terrible, such as taking drugs. Bug-eating, then, is presumably to be regarded as aberrant behavior, with dire consequences. The problem I have with this logic is that bug-eating is an almost universal phenomenon; the U.S. and Europe are curiously alone in their aversion to edible insects. Perusal of some six years of back issues of the *Food Insects Newsletter* as well as correspondence with its founder provides evidence of such behavior in more than 45 countries (including Angola, Australia, Botswana, Brazil, Burma, Cameroon, Chile, China, Colombia, Congo, Egypt, Gabon, Ghana, India, Indonesia, Iran, Ivory Coast, Japan, Kenya, Korea, Laos, Madagascar, Malaysia, Mexico, Morocco, Nigeria, North Africa, Papua New Guinea, Peru, Philippines, Polynesia, Sao Tome and Principe, Senegal, South Africa, Sri Lanka, Tanzania, Thailand, Tunisia, Turkey, Uganda, Vietnam, Zaire, and Zimbabwe).

Although, admittedly the list is not as long as that of countries in which Coca-Cola is sold in cans (93, according to the Coke Can Collectors web page), it's still impressive. Impressive, too, is the fact that there are places in the world, like Angola, Congo, and Laos, where insects are consumed but Coke in cans is unavailable. I guess that means not everything goes better with Coke, after all. According to experts, approximately 500 species of insects in more than 260 genera and 70 families have graced a dinner plate somewhere in the world.

BUZZWORDS

Of course, I feel a little hypocritical criticizing the PDFA for their ad copy, because the truth of the matter is that I am not among the multitudinous ranks of insect-eaters. I've been a vegetarian for more than 23 years, during which time I haven't knowingly eaten anything that moves on its own volition. Normally, being a vegetarian hasn't been a career impediment, but I find that I have a major credibility problem when I give lectures on entomophagy. In the general education entomology course I teach for nonscientists ("Insects and People"), not only is there the lecture, but there's also a laboratory exercise that involves preparing and consuming dishes from around the world that contain insects. Students are quick to notice that I am not a participant in these festivities. And I can't always rely on the teaching assistant to set the example. I thought I was in luck the year that Gwen Fondufe, from Cameroon, was my TA; she had grown up eating stir-fried termites on a regular basis. It turned out, though, that cultural aversions to food are complex; although termites weren't a problem for her, she categorically refused to eat fried waxworms, which she thought were totally repulsive. It's not unreasonable, I guess, given that caterpillars and termites are in entirely different taxonomic orders. Most consumers of mammals who happily eat cows and other even-toed ungulates (order Artiodactyla) might look askance at eating, say, naked mole rats (order Rodentia) or sucker-footed bats (order Chiroptera). Thus, I've been forced on occasion to recruit my longsuffering and incredibly tolerant spouse, Richard, to the effort, bringing him to the class to demonstrate to students that, while I might not consume insects myself, I am willing to give them to a person I hold near and dear. Every year, though, some smart aleck expresses some doubt as to the depth of my affection for this wonderful man.

How the world sees insects

On an intellectual level, I'm completely familiar with the arguments in favor of entomophagy—insects are, after all, a complete source of high quality protein, they're rich in vitamins and minerals, and they come in an almost infinite variety of flavors. For those with deeply felt religious convictions, there's even the consolation of knowing that some of them are certified kosher according to the Old Testament, although I can't quite see smoked locusts ever replacing lox on a bagel. Even etymological arguments are persuasive; as V. M. Holt, a nineteenth-century gastronome, so aptly observed, crickets are especially appropriate as food in "that their very name, *Gryllus,* is in itself an invitation to cook them." But, at the same time, I can emphathize with the insect-averse, including the writers of "Big Ol' Bug."

I wonder, though, if they realize that pushing entomophagy might not exactly be appropriate for their cause. There is no shortage of insects that, if eaten, can hurt your body and your brain, make you sad and your parents mad, and make you clumsy, slow, and numb. One case in point—an AP wire service story from September 11, 1993 reported that

> Even the lowly ant isn't safe from Persian Gulf teen-agers in search of exotic new 'highs.' Adolescents in the free-wheeling port of Dubai are smoking the tiny insects or sniffing the fumes they emit when crushed, the English language *Gulf News* reported. Hameed el-Khafeef of the Dubai police forensic lab was quoted Friday as saying a number of youths were arrested for intoxication after getting high on ants. The practice has become so popular that a small packet of 'Samaseem'—Gulf Arabic for ants—sells for up to $135 in the emirate of Abu Dhabi. . . . The Persian Gulf has been a lucrative market for illicit drugs since the oil boom of the 1970s. But the daily quoted Dubai police as saying youngsters are trying alternative substances either because they can't afford the usual narcotics,

heroin and hashish, or they believe they won't be prosecuted for getting high on ants.

I expect far fewer people here in the U.S. saw this wire service story than saw a PDFA advertisement, and even if news of the psychotropic effects of ants becomes disseminated widely I doubt that bugs will become the new street drug of choice. All the same, it does put a new slant on the idea of an anty-drug campaign.

Is Paris buzzing?

Throughout the United States and elsewhere in the world, festivals are a means for building community pride and potentially for filling local coffers with tourist dollars. A lot of festivals focus on food—among other things, food-related festivals lend themselves well to eating-related activities, always popular pastimes when people gather. Here in Illinois, not unexpectedly, there are quite a few corn-related festivals. There's a Corn Boil in Sugar Grove, Sweetcorn Festivals in Hoopeston, Mendota, and Urbana, and a Popcorn Festival in Casey. Agricultural diversity within the state, though, is higher than most people might imagine, as evidenced by herb festivals in Decatur and Momence, strawberry festivals in Newton, Elmwood, and Kankakee, a peach festival in Cobden, a gooseberry festival in Watseka, and an International Horseradish Festival in Collinsville. Prepared foods also get their day in the sun, even though the wisdom of leaving them there for any length of time might be questionable. There's an Apple Dumpling festival in Atwood, a Cheese Festival in Arthur, a BagelFest in Mattoon, and, in neighboring Indiana, there's a Pierogi Festival in Whiting and a Hot Dog Festival in Frankfort.

Despite the many opportunities available to me, festival-wise, I'd never actually succeeded in attending a festival until 1997. A few years earlier, I went to Collinsville, Illinois, where my mother-in-law lived, and missed the Horseradish Festival by a week, although I did just happen to catch a glimpse of the city's 20-foot-tall inflatable horseradish sitting on the back of a flatbed truck. I'm still kicking myself for missing out on the 1996 Mattoon Bagelfest, where I could have seen the world's record for largest blueberry bagel get shattered before a crowd of thousands. Missing these activities is perhaps excusable, in light of my busy schedule, but there's one festival that I just could never forgive myself for missing. I'm extremely fortunate to live within an hour's drive of one of the nation's few insect-related festivals. Paris, Illinois, is the home of the oldest (if not the only) honey bee festival in the United States.

Now, festivals celebrating insects are few and far between. First of all, there's the food thing—most insect species won't entice hungry out-of-towners into paying a visit and taking a taste. Secondly, most insects aren't quite as user-friendly as fruit, vegetables, or pastries. But such festivals do indeed exist. There are at least two festivals, for example, celebrating woolly bear caterpillars (larvae, in the family Arctiidae). The Woolly Worm festival in Beattyvillle (Lee County), Kentucky, began in 1987 and features, among other things, a Woolly Worm Princess Pageant. But Banner Elk, North Carolina, can boast of hosting the nation's oldest woolly worm festival, which has been held on the third weekend in October every year dating back to 1977 and which features the classic "running of the worms," an intense competition lasting several days (and over 40 heats) to identify the fleetest woolly worm in town.

And noxious pests, perversely enough, have their share of

festivals. Marshall, Texas, hosts an annual fire ant festival and every summer there's a mosquito festival (technically a Mosquito Awareness festival) in Crowley, Arkansas, located in the northeast corner of the state. The Crowley festival features a a World Champion Mosquito Calling Contest as well as a mosquito cookoff, with such dishes as the 1997 third prize-winner, Moosejavian Screamin' Hot Mosquito Wings. Tourists heading for the Cache Valley in Utah will be disappointed to learn that the Randolph City "Mosquito Daze" was discontinued in 1995, after a 5-year run, apparently after exhausting the local population's tolerance for mosquito-related activities (although the nearby town of Providence continues to host Sauerkraut Days in September). For the world travelers, every September since 1982 there has been a Festival of the Phylloxera, in Sant Sadurni D'Anoia, Spain, honoring (or at least recalling) the plant parasite that decimated the local vineyards a hundred years ago. To commemorate the infestation, kids dress up as grapevines and then parade through town, accompanied by masked adults portraying phylloxera and waving sparklers. The highlight of the parade is a bright yellow wheeled papier-mâché plant-sucking bug the size of a small car that periodically spits fire at parade-watchers. This region of Spain is well known for its wineries (the Freixenet winery, among others, is right in town) and evidently much consumption of the local products accompanies the festivities and no doubt plays a pivotal role in convincing adults to dress up and pretend to be plant lice.

Paris is not the only host of a honey bee festival. Latecomers in Hahira, Georgia, conduct a honey bee festival every year, too—but Paris' festival is by far the older of the two. The festival owes its origin to the lobbying efforts of resident Carl Killion, Sr., the state superintendent of apiary inspection who worked long and hard to convince the U. S. government to issue a postage stamp honoring *Apis mellifera*, the western honey bee. When, in October 1980, the stamp was issued, the first-day cover ceremony took place in Paris, in recognition of Killion's untiring efforts. Every year since then, the citizens of Paris have commemorated the event with a honey bee festival. I'd managed to miss the festival for 16 consecutive years but resolutely decided to attend in 1997. For one thing, attending had become a matter of professional pride, but another factor was the news that 1997 was to be the last year that the festival would be run by the Chamber of Commerce, which had just relinquished responsibility for the festival to the Kiwanis Club. Rumor had it that the nature of the festival might change with the change of hands. So, on September 26, a bright and seasonably pleasant day, I went to the festival, accompanied by my amazingly accommodating spouse and my easily bribed 6-year-old daughter, neither of whom would likely have otherwise chosen a honey bee festival as their choice of activity for the day.

At first glance, the honey bee festival, once we arrived, didn't look much different from any other kind of festival here in central Illinois. To eat, there were the same curly fries and elephant ears and lemon shake-ups that show up at all of the local county fairs. Ostrich burgers were exotic but not exactly thematic fare. A major attraction, dominating the south side of the square, was the "Little German in Paris" tent, run by the Kiwanis and featuring German karaoke, sauerkraut, and bratwurst (and fueling the rumors that, left to their own devices, the Kiwanis might indeed be pursuing new and different themes for the festival). Within a few blocks there were arrays of arts and crafts, a flea market, and carnival rides.

Closer inspection, however, revealed the thematic content. Our first hint should have been the lawn geese lining the street into town; in front of almost every house on the main drag was a plastic goose sporting a black and yellow outfit and a pair of antennae on its head. We realized this was not a strange form of vandalism when we spotted lawn geese bee costumes among the arts and craft items being vended on the main square. Also for sale were beeswax sculptures and a variety of honey products. Down the street, the Historical Society had dutifully set up a beekeeping display, somewhat incongruously situated next to a nineteenth century surgical suite, complete with a variety of alarming shiny metal instruments that looked well-designed for inflicting pain. The local post office rightly got into the spirit, too, and was selling commemorative coffee mugs, Frisbees, and hats, although, oddly enough, not stamps and stationery, which were for sale on the main square from a little truck. A car show later that day featured competition in a division strictly for Super Bees, a type of muscle car built back in the sixties and early seventies. These muscle cars are not to be confused with the shiny yellow convertible roadster we saw parked on the east side of the square, from which projected an enormous black and yellow bee some ten feet up in the air.

Although my time in Paris was short, I did the Paris Chamber of Commerce proud—I managed to come home with a Honey Bee festival t-shirt ($15), two Honey Bee Festival hats ($20), a badge from the 1985 honey bee festival ($1.50), and some commemorative stationery ($1.25). I also picked up a free newspaper, a special section produced by the *Paris Beacon-News* staff, detailing all events and activities associated with the 17th Annual Paris Honeybee Festival. It was only when I was back home, reading through the special section at my leisure, that I realized another important function of festivals—they are a vital public outlet for bad puns. In the case of the Paris festival, among the advertisements in the section was an exhortation from the Citizens National Bank of Paris to "Have A Beeautiful Paris Honeybee and Fall Festival." Wood-N-Things advised visitors to "buzz in" to the shop (as did Miss Amelia's Victorian Gift Shoppe, and a notions store called Paragraphs) and the Paris Goodwill promoted a "Honey Bee of a Sale."

I guess some labored jokes are a small price to pay for three days of good will toward an insect species. The puns were all in good taste, even if they were a bit on the strained side. I don't even want to think about the kinds of puns people might have come up with in Frankfort, Indiana, to promote their Hot Dog Festival.

Infield flies and other sporting types

It has not been my experience that there's a natural affinity between insects and sports. If anything, there would seem to be a fundamental incompatibility, given the ability of insects to bring athletic events to a screeching, unscheduled halt. The Chicago White Sox, for example, blamed their 14-7 loss to the Cleveland Indians in Municipal Stadium in Cleveland on August 24, 1982, on infield flies of the entomological kind. Apparently, storms off Lake Erie blew huge numbers of mosquitoes into the stadium during the game, occasioning multiple delays to allow players to spray themselves with repellent. White Sox relief pitcher Jim Kern had particular difficulties and told a reporter, "I felt like I was doing a (bleeping) Off commercial. . . . I couldn't concentrate. They were flying up my nose, in my ears, and I must have swallowed a dozen of them. I got my usual Cleveland greeting, 10,000 (bleeping) flies."

Presumably, these mosquitoes were tormenting players on both sides, and not just the visitors; implied in Kern's comments is the notion that home team Cleveland pitchers are as accustomed to swallowing flies as fielding them. How responsible the flies were for the loss is hard to say. The White Sox had lost seven of the eight games directly preceding their loss in Cleveland and it seems unlikely that insects were at fault in all cases. The point is, though, that baseball players are, by virtue of experience, uniquely entitled to resent insects. After all, baseball games are generally played outdoors—where insects flourish—during the summer—when insects are at their peak. Although football is played outdoors, games generally occur late enough in the year that most self-respecting insects are passing in diapause, enjoying an extended time-out. Basketball is played in indoor arenas during the winter, as is ice hockey—which presents an even greater challenge to insects by virtue of freezing temperatures. Although winter stoneflies, snowfleas, and grylloblattids might be able to cope with the freezing temperatures, the odds are low that they'd be hanging around an ice arena during a game in numbers large enough to provide distraction, particularly to hockey fans.

You'd think, then, that familiarity would breed contempt for insects in professional baseball—but, oddly enough, baseball outdoes all other forms of sport in adopting insects as mascots. Admittedly, it's not major league baseball where these mascots show up. The last major league baseball team named for an arthropod may well have been the 1899 Cleveland Spiders, a team perhaps most distinguished by its remarkable but unenviable season record of 20-134. The team apparently earned the name by virtue of a league executive's remark that the players "look skinny and spindly, just like spiders." The season opened with a 10 to 1 defeat at the hands of their sister team in St. Louis; by June, according to the *Spalding Offical Baseball Guide for 1900*, the team had become the permanent "occupant of the last ditch," 32 games out of first place. Attendance by dispirited fans at home games by the end of the season was so poor that the Spiders were forced to play their last 34 games away (and lost 33 of them).

This dismal example of the consequences of assuming an arthropod identity, though, has done little among the various and sundry minor leagues to discourage the practice at present. The taxonomic diversity of arthropod mascots is actually fairly remarkable. Of 14 teams in the Single A Southeast Division, for example, three are named for insects: the Augusta (Georgia) Greenjackets, the Piedmont (North Carolina) Boll Weevils, and the Savannah (Georgia) Sand Gnats. That number rises to four if all classes of Arthropoda are considered and the Hickory (North Carolina) Crawdads are included. In the Midwest, there are the Burlington (Iowa) Bees. The AAA Pacific Coast League boasts the Salt Lake City Buzz, and in the Southwest there are the Lubbock (Texas) Crickets and the Scottsdale (Arizona) Scorpions.

It's a little difficult to understand why baseball teams opt for an arthropod image. Maybe they see insects as implacable foes, sure to strike fear in the hearts of enemies. Thus, vicious, stinging bees could keep company with Bulls (Durham, North Carolina), Bears (Yakima, Washington), Cobras (Kissimmee, Florida), Timber Rattlers (Wisconsin), Snappers (Beloit, Wisconsin), Sharks (Honolulu, Hawaii), Warthogs (Winston-Salem, North Carolina), and possibly Prairie Dogs (Abeline, Texas) (hey, they can give you a nasty bite, which could get infected and cause big problems!). Or maybe baseball teams consider insects as one of those unstoppable forces of nature that characterize baseball seasons in different parts of the country, like Avalanche (Salem, Oregon), Thunder (Trenton, New Jersey), Cane Fires (Oahu, Hawaii), Quakes (Rancho Cucamonga, California), and Tourists (Asheville, North Carolina). Or maybe there really is a genuine regional fondness for what would elsewhere be objects of indifference at best. It's difficult otherwise to come up with an explanation for the Lansing (Michigan) Lugnuts or the Cedar Rapids (Iowa) Kernels.

For whatever reason they're selected, you'd think arthropods exist in sufficient abundance to provide expansion teams with identities well into the next century. Such, however, doesn't seem to be the case; there's apparently a shortage of arthropods adjudged suitable for mascot service. On April 3, 1998, Georgia Tech University sued the Salt Lake City Buzz, a Minnesota Twins farm team, over the use of the word 'Buzz' and image of what is described as a yellow jacket. According to Mark Smith, executive assistant to the president of the University, the proliferation of Buzzes is confusing to baseball fans and Georgia Tech, which filed for a trademark in 1988, is entitled to exclusive rights: "Buzz is essentially synonymous with Georgia Tech and we need to preserve our image."

This suit did not, of course, sit well with the Salt Lake City Buzz, which immediately filed a countersuit. Owner Joe Buzas, who, by virtue of name alone is entitled to an opinion on the subject, was quick to point out that Salt Lake City had been using the name "Buzz" for three years before Georgia Tech even became aware of the team's existence. That awareness arose when a representative of Georgia Tech spied a Salt Lake City Buzz baseball hat at a trade show in Atlanta. Buzas pointed out that, if in three years Georgia Tech officials failed to become confused, then the potential for confusing alumni in the future was probably minimal. Moreover, according to Buzas, the mascot of the Salt Lake City Team isn't even "Buzz"—it's "Buzzy." Nobody in this discussion, by the way, seems to have noticed that Buzz actually *is* the name of the mascot of the Burlington (Iowa) Bees, who have escaped litigation up to this point.

In response to these arguments, Georgia Tech officials, unmollified and undaunted, pointed out that the mascots are "remarkably similar." I happen to own both Buzz and Georgia Tech baseball caps and can authoritatively say that I don't see the similarity. They resemble each other in the sense that both mascots possess wings, stingers, and antennae, but, then, these features are shared by almost every aculeate hymenopteran and can't really be the source of concern, or else the guys that make Bumblebee brand tuna have grounds for a lawsuit, too. As for colors, which aren't quite as true to nature as is the overall body plan, the Georgia Tech Yellowjacket is definitely navy, alternating with yellow stripes, while the Salt Lake City Buzz bee is more of a cerulean and unadorned by stripes. Moreover, the Georgia Tech Yellowjacket has clenched fists (tarsi?) and an angry grimace (not unlike Buzz, the Burlington, Iowa, Bee), while the Salt Lake City Buzz looks fairly serene and lacks any visible appendages. And the Georgia Tech yellowjacket is the only one of the two wearing gloves.

From what little I know of the law I would think that, if the gloves don't fit, the court is going to have to acquit. As an entomologist, it seems to me that there are about as many differences between these two symbols as there are between a real yellowjacket and a real bee—although, based on my experience with my daughter's first grade teacher, it may be that, to the general public, there are no discernible differences between a real yellowjacket and a real bee. The *Salt Lake Tribune* (April 16, 1998) in fact refers to the Yellow Jacket mascot as a "bee" rather than a yellowjacket.

How the world sees insects

Joe Buzas summarized the crux of the problem best: "What does a bee do? It buzzes. You can't own something that an insect does. . . . We are known all over the country as the Buzz. They are the Yellow Jackets. They are not the Buzz. . . ." (*Salt Lake Tribune,* April 16, 1998). Georgia Tech is proceeding in Federal Court in Georgia, while Salt Lake City has appealed to the U. S. District Court in Denver to get the case moved to Utah. By all accounts, this case will drag on for months.

I could say something like, "Only in America could there be a lawsuit over who owns what a bee does," but I'd only be partly right. It could happen in England and in other English-speaking countries, too, but it couldn't happen in too many other places, because bees don't even buzz everywhere. In Germany, *summen* is what bees do; it's *zumbar* in Spain, *zimzum* in Israel, *anebbebe* in Ethiopia, *wengweng* in China, *bun* in Japan, *brommen* in Holland (but only bumble bees, not honey bees), *bourdonner* in France, and *brzecza* in Poland. In India, there's a veritable beehive of Babel; in Hindi, it's *bhinbhinaanaa*, in Marathi *ghunjan karne* and in Kashmiri *gīīgīī karun*. Any one of these terms would be a great name for a baseball team, although rousing cheers might be difficult to inspire if no one in the stands could pronounce it properly and it might be hard to print diacritical marks on a baseball cap. I guess if there's a lesson here, it's that buzzwords really do matter to people, a sentiment I've maintained all along.

Sounding off

’m often invited to give public lectures on entomological topics and, always anxious to do my part for science literacy, I rarely refuse such invitations. One such invitation came from the University of Illinois Alumni Foundation, to speak to a group of alumni attending a picnic at a park not far from campus. For my services, the Foundation was prepared to offer me reimbursement for my expenses and all the potato salad I could eat; times being what they were, I jumped at the offer. I found myself seated at the end of a long table surrounded by total strangers with nothing apparent in common other than having attended classes at the University of Illinois at some point during the last century. When the man seated next to me discovered that I was the evening's featured speaker, he got very excited, exclaimed, "Then you must know all about this!", plunged his hands under the table, and began wrestling with his belt. Much to my enormous relief, all he did was to produce a tubular device that resembled a small flashlight and show it off to me triumphantly. Seeing my blank stare, he proceeded to inform me, with not a little disappointment evident in his voice, that it was an electronic mosquito repeller.

How the world sees insects

My entomological credentials obviously rendered suspect at this point, he diplomatically asked me to pass the pickle relish.

Up until that moment, I had never seen an electronic mosquito repeller. In fact, I thought that the principle on which these devices operate, namely that mosquitoes are repelled by certain types of sound, was long ago discredited. Thus, I was amazed to find that a quick search of mail order catalogues at home (of which we invariably have a huge supply) turned up dozens of these devices for sale. They run the gamut from sublime to ridiculous. The Cadillac of electronic mosquito repellers must surely be the Moltron III, the "Swiss-proven bug shield" advertised by the Sharper Image catalogue. According to the catalogue, "The Moltron bug shield protects you by electronically recreating the sound of a *male* mosquito, which pregnant females hate." The Moltron III (Moltrons I and II must have never left the drawing board) lists for $19.95 and is the only one of these devices with an adjustable dial, instructions in three languages (English, Spanish, and French), and a disclaimer ("since there are at least 67 known species in the U.S. alone, some are bound to fall outside the fixed range"). The manufacturers modestly "hope you find your Moltron III to be effective as most people do," an ambiguous wish at best. Among other things, it's mostly inseminated females with as yet undeveloped eggs that actively seek out blood meals— pregnant ones are busy seeking out places to lay their eggs and don't as a rule crave blood meals or pickles or any other unusual food items. It's a pretty safe bet that virgins aren't quite so repelled by the sound of male mosquitoes, be it live or on tape, or else there wouldn't be so many mosquitoes around to repel.

"Bye! Bye! Mosquitos" operates on the same principle as does the Moltron III, but does so on a greatly reduced budget—at $4.98, it's more like the Yugo of electronic mosquito repellers. This one is from China (as opposed to "Swiss-proven") and the instructions are in only one language, which bears a passing resemblance to English:

> Mosquitos frequently infect you place in Summer, especially at night. They are extremely irritating as they disturbed our sleep and the mosst annoying of all is the difficulty in getting rid of the itch & soreness. After, ordinary mosquito-increase or "Electrified Mosquito Killer" are used. However the odour is unbearable and the abuse of some of them may become dangerous. In order to do away with the above nuisance, a brand new production called "Electronic Mosquito Repeller" Has now een produced. According to the research of insect ecology, most of biting mosquitos are female ones in spawning period. A Spawning female mosquito is very disgusted at the approaching of male mosquito. There-fore, the trequency of Repel-It is made to imitate the sound signal of male mosquitos to repell female mosquitos away.

This one doesn't even come with the AA battery it requires for operation.

"Bye! Bye! Mosquito" is available through the Harriet Carter catalogue, a veritable treasure trove of electronic mosquito repellers. In one catalogue, sandwiched in among the scented toilet tissue spindles, invisible tummy trimmers, cordless electric safety trimmers ("neatly removes unsightly nose, ear, and brow hair!") and "Cat-cans" ("Train your cat to use a toilet!") are the ultrasonic necklace mosquito repeller ($6.98), the clip-on "Bye! Bye! Mosquito," and the "Mosquito Chaser" ($7.98), a boxlike device with a "nylon cord to hang the unit from a convenient location." Whereas the clip-on is from China, the necklace and the box come from Taiwan. The manufacturers of the necklace state cryptically that the device "chases mosquitos away by transmitting a barely audible sound wave that mosquitos hate." Interestingly, the instructions do warn that while "the sonic frequencies generated by this unit are harmless to humans and pets . . . discontinue using the unit near persons who are overly sensitive to sound." Evidently, it's not just mosquitoes who are repelled by these things. As for "barely audible," I suppose everything is relative but I can see somebody who is borderline psychotic being driven right over the edge after an hour or two of having one of these hanging around his neck. By the way, the Mosquito Chaser doesn't have an "off" switch.

There's at least one other product on the market for skeptics who don't believe that spawning mosquitoes can't stand to hear their mates whine—the Love Bug ($12.95, Hand-in-Hand catalogue) "repels mosquitoes by electronically duplicating the wingbeat of the dragonfly—the mosquito's mortal enemy!" The ad shows this device clipped to a baby's stroller and states that it "actually keeps mosquitoes away within a 20 to 30 foot radius with the same degree of effectiveness as the best lotions and sprays." I suppose they must mean they're as effective as the best suntan lotions and oven cleaner sprays, since the manufacturers don't specify which dragonfly is the one species of dragonfly whose wingbeats are electronically duplicated and it seems unlikely that all species of mosquitoes in all 50 states (and possibly Guam and Puerto Rico) will recognize the wingbeats of the one dragonfly species chosen to represent the entire order Odonata.

The manufacturers of these devices—you can just imagine them, consulting with Swiss scientists and Chinese researchers in insect ecology—seem to possess information that is not available in the refereed scientific literature. The only references I could

find about the efficacy of electronic mosquito repellents suggested strongly that there isn't any. If these devices aren't merely unscrupulous and potentially dangerous consumer rip-offs, then we Americans have some major catching up to do. If indeed mosquitoes can be repelled by ecologically significant sounds, the possibilities for designing new electronic repellents are virtually endless. Tests should begin immediately with devices that simulate, say, the sound of a hand slapping and crushing an engorged female. Or the sound of an aerosal can of mosquito repellent being discharged, a control technique that doesn't even hurt the ozone layer. Or maybe a device that plays the soundtrack from the film, "The Birds," or one that repeatedly announces that the ban on domestic DDT use has been lifted. The federal government may even be in a position to facilitate such research. After all, they probably still have those rock and roll tapes they used in Panama to drive Manuel Noriega out of the Papal Nuncio's residence—who knows, maybe they'll work just as well on female mosquitoes?

How
entomologists
see
themselves

Entomological legwork

While searching through a bookcase one day for some otherwise forgettable reference, I came across my collection of *Hexapod Herald*s. The *Hexapod Herald* was a newsletter produced intermittently by the entomology graduate students at the University of Illinois between 1983 and 1986. An eclectic publication, it contained insect-related news, crossword puzzles, recipes, poems, jokes and riddles, and other sundries. In 1986, then-editor James Nitao published the results of a survey he had conducted among the faculty and students here at the time. Questions dealt with career interests and development, as well as personal interests and hobbies. When questioned about their early childhood experiences, 12 of 15 faculty respondents admitted to having made an insect collection, 3 to having had an ant farm, 3 to having burned ants with a magnifying glass and 6 (40%) to having pulled wings off flies. The graduate students weren't much better; of 31 graduate students, 12 had burned ants with a magnifying glass and fully 17 (55%) admitted to having pulled wings off flies. Among other activities reported by respondents were "spider fights, pulling legs off daddy longlegs, terrorizing . . . lightning bugs, blowing up ants with caps, tying string around horse flies"

BUZZWORDS

and doing things with pieces of straw in the vicinity of the nether ends of flies that, in the interest of decency, will not be detailed further here.

Such information gives one pause, certainly as to whether entomologists are atypical with regard to this sort of behavior. According to Vincent Dethier, in his book "To Know a Fly"

> . . . there never seemed to be a taboo against pulling off the legs or wings of flies. Most children eventually outgrow this behavior. Those who do not either come to a bad end or become biologists.

To Dethier, the distinction between those two outcomes was clear; I suspect that, in the view of the general public, they may appear as one and the same.

The Department of Entomology at the University of Illinois has a long and illustrious history of removing appendages from insects in the name of science. As early as 1932, before his arrival here, Gottfried Fraenkel clipped off the halteres of flies (the balancing organs behind the sole pair of wings) to see what would happen to their flight responses. By doing so, he became part of a rich entomological tradition dating back more than two centuries, to 1716, when W. Derham, rector of the Upminster Church, reported in *Physico-Theology* that "if both be cut off, they [flies] will fly awkwardly and unsteady, manifesting the defect of some very necessary part." A current member of our faculty, Dr. Fred Delcomyn, does not pull legs, wings, or halteres off flies but in the course of twenty years of study of insect locomotion he has pulled the legs off innumerable cockroaches and thus is something of an authority on the subject. He has traced the scientific pulling-off of insect legs back to 1888, when G. Carlet published a paper in the *Comptes Rendu de L'Academie de Science* entitled, "*De la marche d'un insecte rendu tetrapode par la suppression d'une paire de pattes.*" Dr. Delcomyn is a cut above most of his colleagues that remove appendages in that he is one of few who has pioneered the design and use of artificial limbs for cockroaches (picking up on a general practice evidently started by one W. von Buddenbrock more than seventy years ago). Dr. Delcomyn's interest in cockroach prosthetic limbs is purely academic; the cockroach prosthetic limb is simply not destined for success in the consumer mass market.

That there is a rich literature on the effects of removing legs or wings on insect locomotory behavior is not that surprising, and I guess it's also not surprising that this literature is substantially larger than the body of literature addressing the question as to whether insects feel pain. In fact, I was able to find only two papers dealing with the subject: one, by V.B. Wigglesworth entitled "Do insects feel pain?" published in 1980 and one by C.H. Eisemann et al., entitled "Do insects feel pain?—A biological view" written four years later. Authors of both papers conclude that, although it's difficult to say definitively, it's likely that insects don't feel pain as we define it. One relevant example of the sort of evidence used to support this conclusion is the observation that insects do not show "protective behavior towards injured body parts, such as limping after leg injury. . . . On the contrary, our experience has been that insects will continue with normal activities even after severe injury or removal of body parts. An insect walking with a crushed tarsus, for example, will continue applying it to the substrate with undiminished force" (Eisemann et al., p. 166). In contrast, injuries far short of crushing can reduce adult humans to a state of abject whimpering—an ingrown toenail, for example, sent my husband to a hospital emergency room not long ago.

BUZZWORDS

My own opinion on insect pain responses was pretty much solidified by an experiment I conducted as a graduate student in Henry Hagedorn's insect physiology class at Cornell University in 1980. In our exercise on "the insect heart," we were instructed to anesthetize a cockroach, pin it dorsum-down in a dissecting dish and then, "working rapidly but with precision," cut off the head and legs, remove the ventral body wall, clear out the visceral mass, scrape out the fat body, and expose the dorsal vessel. We then pumped in cold saline to observe the effects on heartbeat as the temperature dropped 20 degrees; and we dripped in various physiological salines to determine whether or not they would stop the heart from beating altogether. Anyone whose cockroach heart was still beating at this point was encouraged to test the effects of various neuroactive substances, including nicotine, acetylcholine, and caffeine, on heartbeat. After about 2-1/2 hours of this sort of thing, my laboratory partner, an undergraduate named Steve Passoa, and I had completed all of our assigned tasks. Steve then removed the pins securing the truncated roach in place and, much to my unspeakable horror, the roach remains proceeded to SWIM AWAY, little stumps flailing frantically in the saline. At that moment, I reached the profound realization that cockroaches are just not like us.

So I'm not going to pursue the issue further, having satisfied myself that, if insects feel pain, it's in a way that I can't possibly hope to relate to. But there is one other issue that has bothered me. In April 1994, I was a guest on Whad'ya Know, a popular quiz program on Wisconsin Public Radio. I was prepared for most of the questions, of the usual sort (e.g., "What good are mosquitoes?") but host Michael Feldman truly threw me for a loop when he asked me whether or not insects have free will. Although this question was evidently a burning issue around the turn of the twentieth century, a computer search of the recent literature with the key words "free will" and "insects" failed to turn up anything useful. Any and all suggestions are appreciated.

"What's in a name? That which you call Eltringham's gland. . . . "

Every now and then, I get mail addressed to me care of the Department of Etymology. It's an understandable mistake—neither entomology nor etymology is widely known as a profession by the public at large and the lexicographic similarity is certainly striking. I also have received letters addressed to the Department of Antomology, yet another understandable mistake, and I once even received one addressed to the Department of Endocrinology. Etymology, of course, is the study of word origins, and entomology is the study of insects. Every now and then, however, entomologists become etymologically involved with their subject, particularly when insect body parts acquire proper names.

Eponymous body parts abound in human anatomy and physiology—the human body is a veritable football roster of names. There are cells, glands, ligaments, capsules, loops, ducts, tubes, nodes, apparati, and canals named in honor of the men who first described them (Purkinje, Gley, Berry, Bowman, Henle, Mueller, Eustachios, Ranvier, Golgi, and Havers, respectively). This penchant for naming things after people probably stems from human fascination with exploring the unknown—it's undoubtedly why continents and mountain peaks are named for their discoverers. It's only a small logical jump for homebound physicians to compare their anatomical adventures with the more traditional sort, leading perhaps to travelogues such as "On navigating the islets of Langerhans," "Boating the Haversian canals" or "Sailing around the angle of Louis." The angle of Louis, by the way, is the angle made by the sternum at the second intercostal space in the rib cage. It's named in honor of Antoine Louis, 1723-1792, whose greater fame may have been as co-inventor along with J. I. Guillotin of an eponymous instrument of law enforcement.

Entomologists are hardly immune from this quest for immortality. Eponymous insect parts abound. Unfortunately, very few of these insect parts gain a reputation outside a rather narrow range of specialists. The great Malpighi, professor of anatomy at several universities, as well as physician to Pope Innocent XII, hit it big with his silkworm dissections back in the seventeenth century. The tubules he described, ultimately called Malpighian tubules, are kidney-like organs found in just about every insect species, except for aphids and a few other nonconformists. Christopher Johnston, a physician by training like Malpighi, lucked out in 1855, when he described the "auditory capsule" at the base of the antennae of mosquitoes; he probably had no idea that Johnston's organs are widely distributed among insects. Now, not only mosquito experts know of their existence but generations of students have had to learn about them to prepare for quiz questions in insect anatomy classes.

More typically, though, having an insect body part named in your honor is hardly a shortcut to lasting fame. Consider H. Eltringham, who has not one but two glandular structures named in his honor in neuropterans: an extrusible abdominal gland in the

BUZZWORDS

mantisfly, *Mantispa styriaca,* and a scent gland in the hindwing of the antlion, *Myrmeleon nostras.* Not even the great Malpighi can boast of two insect organs named in his honor, yet where has it gotten Eltringham? For that matter, how many people, other than Snodgrass and Imms, long-dead authors of insect anatomy texts, know where to find Weismann's ring, Semper's rib, Latreille's segment, Dufour's gland, or the organs of Tomosvary, Gabe, Hanstrom, Schneider, or Nabert?

What's even more depressing, from the perspective of lasting fame, is that knowing where an organ is doesn't even guarantee that anyone will recognize its namesake. An informal poll of my colleagues here at the University of Illinois, inquiring as to whether anyone knew for whom the glands of Philippi were named (accessory glands associated with silk glands of caterpillars), invariably elicited the less than helpful response, "Some guy named Philippi?" I have no idea for whom Hicks' bottles—which, according to Snodgrass are "flask-shaped pits or depressions in the antennae of bees or ants"—were named, and whether or not this Hicks had some kind of drinking problem.

These matters are hardly esoteric and abstruse. In case you missed it while thumbing through your back issues of *Memoires de Biospeologie,* a journal dedicated to the biology of cave-dwelling organisms, the organ of Bellonci was described for the first time in stenasellid isopods. This organ is known from several crustacean orders but Pitzalis et al. (1991) were the first to have spotted them lurking "near the rostral corner of the cephalon," in stenasellids. I can't help thinking that Bellonci would be proud. I also can't help wondering who the heck Bellonci was.

I did try to find out, but I was completely unsuccessful. Bellonci's name doesn't appear in the *Compendium of the Biographi-*

cal Literature on Deceased Entomologists. Yes, there is such a book, compiled by Pamela Gilbert, Entomology Librarian at the British Museum. Of course Bellonci's not in the book, you say with righteous indignation, he described organs in crustaceans, which aren't insects, so why should he be listed in a compendium of entomologists? But where dead entomologists are concerned, Pamela Gilbert is quite a liberal, stating that, as a matter of policy, "It has seemed to me more useful to be embracive rather than restrictive. In fact one cannot be pedantic about defining entomology." Even with such a broad view, Bellonci's is not among the 7,500 names in the compendium. I suppose another reason his name isn't listed is that he might not be dead yet. The criteria for defining "deceased" are a lot less ambiguous, one assumes, than the criteria used for defining "entomologists."

There is one odd similarity between medical and entomological eponyms: an extraordinarily high proportion of eponymous body parts seem to be concentrated in reproductive organs. The human (particularly female) reproductive system is fairly burgeoning with physicians vying for their piece of immortality. Fallopio found tubes, Bartholin described glands, and de Graaf discovered follicles. One might even count J. Braxton Hicks in this number. J. Braxton Hicks wrote a long series of papers in the *Transactions of the Linnaean Society of London* between 1853 and 1859 describing various structures of antennae and wings of species in many insect orders. Entomology was for Braxton Hicks, however, simply an avocational pursuit; by day he practiced as an obstetrician and gained immortality by describing uterine contractions of false labor during pregnancy.

Among insects, male entomologists have left their mark on the reproductive structures of female insects. The peculiar tissue mass in female bedbugs that absorbs the force of the incoming intromittent organ of the male during the process understatedly known as "traumatic insemination" is known not by one but by two eponymous names. It's variously called the organ of Ribaga and the organ of Berlese. The latter name is a reference to Berlese, who also earned eponymous immortality among entomologists by designing a funnel for capturing soil-dwelling arthropods. Why (male) entomologists would like forever to have their names associated with a structure used during bizarre copulation by a bloodsucking and ill-smelling parasite is beyond my comprehension. The only explanation I can think of, which I guess would apply equally well to both human and insect anatomists, is this notion of anatomist as explorer in unknown terrain. Since the vast majority of these eponymous explorers are or were male, one supposes the female reproductive tract may very well seem enigmatic and possibly fraught with hidden dangers. All this notwithstanding, it still sounds a little funny to me. After all, how did Mrs. Berlese or Mrs. Ribaga feel about having their husbands' organs prominently displayed in entomology textbooks for all to see?

Apis, Apis, Bobapis

I've never been particularly gifted at naming things. To illustrate, for many years I lived with two cats whom I adopted and named; one was called, logically if not terribly imaginatively, "Pussins" and the other was called, imaginatively if not terribly logically, "Nooners." Pussins and Nooners are not the only pets I've had occasion to name; in second grade, I had a series of red-eared turtles as pets—the type sold in pet stores in little plastic bowls equippped with a green plastic palm tree projecting from the center island. As I recall, they all had names starting with T (Tommy, Terry, Timmy, Teddy, etc.) and none ever lived longer than three weeks—so I guess I'm also not particularly gifted at turtle-rearing.

This problem I have with coming up with names is the reason I have so much respect and admiration for systematists who must, as a matter of course, invent names for species on a routine basis. These names, unlike the names of my turtles, must last longer than three weeks—theoretically, they're supposed to last forever. Fortunately for me, I guess, I rarely encounter, much less describe, new species in my line of entomology. I did once find a species of weevil in the genus *Apion* that, according to the expert at the

Smithsonian to whom I sent it for identification, was undescribed. I still live in fear that when it is described for posterity by a systematist in a refereed scientific publication it will be named after me. The last thing I need is for a beetle whose distinguishing feature is a proboscis fully half the length of its body to be known as "Berenbaum's weevil."

But systematists find new species all the time. It's not that Linnaeus didn't do a spectacular job naming all living things, but there are many thousands more species needing names today than were recognized two centuries ago. Today there are probably more species of *Apion* than there were beetles with Linnaean names back in 1758. Insect names run the gamut literally from A (*Aaages*, a beetle described by Barovksii in 1926) to Z (*Zyzzyva*, another beetle, described by Casey in 1922). With so many species to name, it's not unreasonable that systematists, particularly entomological systematists, occasionally get tapped out. W.D. Kearfoot, for example, described a series of species of moths in the genus *Eucosma* in 1907 and gave them rhyming names running through most of the consonants in the alphabet, including *bobana, cocana, dodana, fofana, hohana, kokana, lolana, momana, popana, rorana, sosana, totana,* and *vovana* (anticipating by 60 years the song, "The Name Game," by Shirley Ellis—you know, "if the first two letters are ever the same, you drop them both then say the name, like Fred Fred drop the F's Fo-red . . . "). Scattered in amongst these are *fandana, gandana, handana, kandana, mandana, nandana, pandana, randana, sandana, tandana, vandana, wandana, xandana, yandana,* and *zandana*. To relieve the monotony, he also described a few other species with the more distinctive epithets *boxeana, canariana, floridana, idahoana, miscana, nomana, sonomana, vomonana,* and *womonana*, as well as, inexplicably, *subinvicta*.

How entomologists see themselves

Most systematists rise to the challenge more creatively than did Kearfoot. Some in fact are so creative that they end up being criticized by the International Zoological Congress. This is the body that drew up a set of standardized rules for naming things and that has continued to meet and to issue guidelines and opinions since 1901. Reading these rules has also convinced me that I was not destined to become a systematist—these rules are about as clear and simple to me as the instructions for filing an income tax statement. It didn't really help matters that the rules are written with alternate pages in French and English; nor was the 47-page bilingual glossary much help, except to keep me up nights wondering why the French glossary is 2-1/2 pages longer than the English one. Is there maybe a naughty section that hasn't been translated?

There have been several systematists who provoked the ire of this body by being a little too creative in their nomenclatural efforts. For example, G.W. Kirkaldy was criticized for frivolity by the Zoological Society of London in 1912 by unobtrusively bestowing upon a series of hemipterans, or true bugs, the generic names *Ochisme, Polychisme, Nanichisme, Marichisme, Dolichisme,* and *Florichisme.* Presumably, eight years elapsed before anyone in the Zoological Society actually pronounced these names out loud and realized that the series provided a plea for osculatory adventures. V.S.L. Pate slipped one by the censors in 1947 when he described a new genus of tiphiid wasp, *Lalapa,* and then proceeded to name the sole species in the genus *Lalapa lusa.* Dr. Arnold Menke of the Systematic Entomology Laboratory at the U.S. National Museum erected the genus *Aha* and proceeded to describe the nominate species *Aha ha.* Such actions have prompted A. Maitland Emmet, in his book, *The Scientific Names of British Lepidoptera: Their History and Meaning*, to declare, "Scientific names have much in common with crossword puzzles. The nomenclator is the setter . . . if he can mystify his fellow entomologists, he will derive sadistic pleasure in so doing."

In some instances, mystification was clearly not the goal—as when B. Neumoegen (1893) in his "Description of a peculiar new liparid genus from Maine," dedicated the genus to his "faithful co-labourer and friend Mr. H.G. Dyar" in naming it the euphonious *Dyaria* (say it out loud with the emphasis on the penultimate syllable). Such actions may well have led to the inclusion in the "Recommendations on the Formation of Names" (Appendix D.I.5) the statement "A zoologist should not propose a name that, when spoken, suggests a bizarre, comical, or otherwise objectionable meaning" (p. 193).

Notwithstanding, systematists have managed to sneak quite a few bizarre and comical names past the censors. The aforementioned Dr. Menke is an absolute master of names; Rumpelstiltskin wouldn't last ten minutes with him. He sent me his own list of more than 100 peculiar scientific names (published in the journal B.O.G.U.S. (Biological and Other Generally Unsupported Statements) 2:24-27 (April Fool's Day, 1993). Dr. Menke's list convinced me that I shouldn't be writing humorous essays at all—the people who came up with some of these names would do a much better job. For example, there's *Townesilitus*, a braconid wasp; *Agra vation*, a carabid beetle; *Castanea inca dincado*, a moth; *La cucaracha* and *La paloma*, two more moths; *Chrysops balzaphire*, a horse fly; *Colon rectum*, a beetle; *Heerz lukenatcha*, *Heerz tooya*, *Panama canalia*, and *Verae peculya*, wasps in the family Braconidae; *Leonardo davincii*, a pyralid moth; *Phthiria relativitae*, a bee fly; *Pison eyvae*, a wasp; *Tabanus nippontucki* and *Tabanus rhizonshine*, both horse flies; and, of course, *Ytu brutus*, a beetle. I don't know for certain, but I'd be willing to bet large sums of money that, even as a child, Dr. Menke would not have been one to saddle a turtle with such a pedestrian name as "Tommy."

So it's a thin line that systematists must watch (and occasionally wink at); thanks to their efforts, scientific names can make for entertaining reading. In particular, names with cultural relevance are, in my estimation, particularly entertaining. This essay was in fact inspired by a letter I received from Dr. Margaret Novak, a water program specialist at the New York State Department of Environmental Conservation. She thought a column on "bizarre and cryptic species names" might be interesting and, to get me started, sent me a photocopy of page 88 of J.H. Epler's 1987 "Revision of the Nearctic *Dicrotendipes* Kieffer 1913 (Diptera: Chironomidae)." For those unfamiliar with the work, page 88 contains a description of the new midge species *Dicrotendipes thanatogratus*, "from the Greek *thanatos*, meaning death, dead; and the Latin *gratus*, meaning thankful, grateful. This species is named for the Grateful Dead, a group of musicians who for the past 20 years have provided the background music for my life." I was subsequently inspired to search for other cultural references among arthropod names and had a few successes. In *A Prehistory of the Far Side*, by Gary Larson, there is reprinted a letter from Dr. Dale Clayton to the cartoonist. In the letter, Clayton proposed naming a new species of owl louse *Strigiphilus garylarsoni*, in recognition of the "enormous contribution that my colleagues and I feel you have made to biology through your cartoons."

Not all cultural references are to people, though. When Jill Yager and colleagues discovered the world's largest remipede crustaceans in caves in the Bahamas in 1986, they were inspired to

Deadhead midge

name the family Godzilliidae, and the nominate genus *Godzillius*, in honor of the largest reptile to rise out of the sea in recent film history. Three years later, when called upon to describe a new genus in the family, Yager rose to the challenge and named it *Pleomothra*; "in keeping with the spirit of the first described godzilliid, the name is derived from the Japanese horror film star 'Mothra' and the Greek word 'pleo,' meaning 'swim.'" I feel compelled to point out, though, that there may be some redundancy here—in the original film, Mothra does indeed swim (as a caterpillar) from Monster Island to Tokyo when the two little girls with whom she communicates telepathically are captured by ruthless entrepreneurs—but that's another story.

The one problem that might arise with culturally referential names is that sometimes cultural values can change. There is, for instance, an extinct palyodictyopteran fossil species described in 1934 by one P. Guthörl as *Rochlingia hitleri*, in honor of a rising political star of the era. A subsequent attempt to synonomize the genus with an older one and rename the species *Scepasma europea* was made by Hermann Haupt in 1949, declaring *R. hitleri* to be a nomen nudum (probably one of the nicer things Hitler has been called), but, according to my colleague Dr. Ellis MacLeod, Haupt's interpretation of the Rules is probably incorrect and Hitler's paleodictyoperan is "at least available if not valid." Dr. MacLeod in all the years I know him never exhibited any neo-Nazi or white supremacist views, so I am confident that his analysis was based on solid nomenclatural grounds.

A less egregious example of how cultural values can leave names in the lurch was described in the *National Enquirer* from February 25, 1992, in an article entitled "A bug named Bush?"

BUZZWORDS

> Scientists want you to name newly discovered species after a beloved person or an enemy! And it's all to benefit nature. Leading scientists who classify new organisms want to raise save-the-habitat funds by auctioning rights to name new species of flowers, birds, bugs, or fish. Recently, a Costa Rican wasp was named *Eurga Gutfreundi* after a disgraced Wall Street trader who ripped off the federal securities market.

While a literature search did indeed confirm the existence of John Gutfreund, Wall Street ripoff artist, I could not at first confirm the existence of this wasp. This would not be the first story published by the *Enquirer* that proved to be difficult to confirm. However, Dr. David Wahl, of the American Entomological Institute of Gainesville, Florida, upon hearing of my difficulties, pointed out that my failure to confirm the existence of a Costa Rican wasp named for Gutfreund stemmed from the fact that the genus name was misspelled in the *National Enquirer* article I had cited. *Eruga* (not *Eurga*) *gutfreundi* is a pimpline ichneumonid in the tribe Polysphinctini, described by I.D. Gauld in 1991. Which means that, except for spelling, the *National Enquirer* story was essentially correct. Which means I'm going to have to start rethinking other entomological stories I see in the tabloids. I guess this means that the two-page story in the November 30, 1993 *Weekly World News* is true—that "2-inch fireflies" that "pack a 600-volt sting" really have "killed dozens of hapless citizens in Central America and Mexico in the past two years" ("they can flatten a grown man with a single jolt" and have been steadily advancing on the U.S. border since "200,000 of them escaped from a top-secret research laboratory in Managua, Nicaragua." The article states that, at the rate they're moving, they "could reach the U.S. border by March." I don't think they've arrived yet, but I guess I better warn Elvis anyway next time I see him at the 7-Eleven.

Department of
Ant-omology?

From time to time, I am reminded of my first day in graduate
school. My soon-to-be thesis advisor, Paul Feeny at Cornell
University, was kind enough to pick me up in his Audi Fox at the
graduate dormitory and drive me out the 3/4 mile or so to his
laboratory. His laboratory was not in either of the buildings
housing most of the entomologists on campus—Comstock or
Caldwell Hall. Instead, he operated out of a ramshackle (borderline decrepit) building called the Insectary. As we pulled up to
park, I noticed that the words "Entomology and Limnology" were
emblazoned on the front of the building. Seeing my quizzical
look, Feeny remarked offhandedly, "They dropped the Limnology
part long ago," and we walked on in.

To this day, I still don't see entomology and limnology as
disciplines logically housed under the same (crumbling) roof.
Limnology, from the Greek *limnos* ("pool," "marsh" or "lake") is
the study of lakes; entomology is the study of insects. Of course,
insects frequently are found in lakes. Then again, insects are
probably even more frequently found in boxes of cereal or lurking
under seat cushions, hardly a justification for, say, a Department of

Entomology and Food Science, or a Department of Entomology and Furniture Studies.

I was most recently reminded of that fateful day in 1993, when various plans for reorganizing the College of Liberal Arts and Sciences and the College of Agriculture were being passed around the University of Illinois campus. Our department, which at eight full-time faculty members barely lost to the Program for Religious Studies for the dubious distinction of being the smallest unit in the College of Liberal Arts and Sciences, figured prominently in many of these schemes. We were encouraged, among other things,

—to merge with Plant Biology to form a Department of Plant Biology and Entomology

—to merge with the Department of Ecology, Ethology, and Evolution, presumably to form a Department of Ecology, Ethology, Evolution, and Entomology (the Departments of English, Economics, and Electrical Engineering declining to participate in this obviously alphabetically motivated move)

—to merge with the Office of Agricultural Entomology in the College of Agriculture, which was itself in the process of considering a merger with several other departments to form a Department of Natural Resources or possibly a Department of Plant Protection

—to buy one-way tickets to Borneo for each of the eight full-time faculty so they can stay and collect butterflies, thus ceasing to cause problems for people in other life science units.

With the exception of the eight aforementioned full-time faculty members, practically nobody thought that leaving us alone to remain a free-standing Department of Entomology was an option worth pursuing.

While pondering our possible future, I conducted an informal survey of entomology programs in the United States by turning to the back pages of the 1992 Entomological Society of America membership directory, in which were listed addresses for most entomology programs in the country. On that list were addresses for 40 Departments of Entomology. Also on the list was one Division of Entomology (University of Idaho), and a Center for Studies in Entomology (Florida A&M). At other institutions, entomology shared billing with a diverse array of disciplines, as in the Department of Entomology and Nematology (University of Florida), the Department of Entomology and Applied Ecology (Delaware), the Department of Agronomy, Horticulture and Entomology (Texas Tech), the Department of Entomology and Plant Pathology (Tennessee), the Department of Entomology, Plant Pathology, and Weed Science (New Mexico State), the Department of Plant, Soil and Insect Science (Wyoming), and, rounding out the list, the Department of Zoology and Entomology (Colorado State). Entomology was also less obviously housed in three Departments of Biology, three Departments of Plant Sciences, a Crop Protection Department, two Plant and Soil Science Departments, and a Department of Zoology.

The University of Illinois, and I say this without local chauvinism, probably leads the nation in confusion with respect to housing entomologists. On our campus at the time reorganization discussions were under way, there was a Department of Entomology housed in the College of Liberal Arts and Sciences, an Office of Agricultural Entomology housed in the College of Agriculture, and a Center for Economic Entomology in the Illinois Natural History Survey (which is actually an autonomous state institution, independent from, but intellectually closely tied to, the university). There were also three entomologists in the Center for Biodiversity at the Natural History Survey, at least one entomologist in the College of Veterinary Medicine, and rumor was that there was an entomologist cleverly concealed in the Department of Urban and Regional Planning in the College of Fine and Applied Arts.

So, why is it so hard for university administrators to find a place to keep their entomologists? It's obviously not that entomologists are regarded as pariahs by the rest of the scientific community—otherwise there would be 50 free-standing departments of entomology, and applied ecologists, plant scientists, and nematologists wouldn't be professing their solidarity with entomologists. In my opinion, the problem may lie in the difficulties people have in finding a place to put insects. It's not always easy even to recognize what belongs in the class Insecta, much less to know where within the vast reaches of the class to place it. Considering that there are about a million species of insects, it's really not all that surprising that people have had a hard time over the centuries trying to figure out where to put them—after all, finding a place to keep a million of anything in order is a challenge.

Taxonomists in particular have grappled with this problem for centuries. Around 1230 A.D., for example, one Bartholomaeus Anglicus authored *De Proprietatibus Rerum*, a 19-volume compendium intended to serve as a complete description of the universe. Insects appear in several places throughout the opus. In Book 12, for example, creatures of the air are featured and flying insects are lumped in with birds. Bees are rather poetically described, along with birds, as "ornaments of the heavens." Book 18 features terrestrial animals and considers collectively "worms, adders and serpents." Even so, Bartholomaeus recognized that insects didn't fit obligingly into the general scheme. The bee, for example, "is a little short beast with many feet. And though he might be classified among flying creatures, yet he uses his feet so much that he can reasonably be considered among ground-going animals."

Bartholomaeus Anglicus can be forgiven for his confusion about the placement of bees in particular and insects in general, since, after all, he was working in the depths of the Dark Ages—the word "insect" hadn't even been coined yet. But even the Scientific Revolution did little to ease the task of finding rightful places in the world for all insect species. The notable taxonomist Schiffermüller, for example, missed by a mile when he described *Papilio coccajus* in 1776; the species he thought was a butterfly and thus confidently placed in the order Lepidoptera, along with other butterflies and moths, was in actuality an ascalaphid neuropteran, or owlfly, an entirely different sort of animal altogether.

Things didn't improve much in the nineteenth century, either. One notable classification that was definitely out of order involved the immature stages of the syrphid fly genus *Microdon*. Adult *Microdon* are perfectly normal looking self-respecting hover flies in the family Syrphidae. The larvae are quite another story. Maggots of *Microdon* are legless, armored, bizarrely ornamented but otherwise featureless little creatures. This body build is ideal for fending off the stings and bites of enraged ants—to which *Microdon* larvae are often subjected by virtue of the fact that their habitat of choice is ant nests—but it is less than revealing for systematists seeking out resemblances to other known life forms. Thus, it is not surprising that these creatures were early on described as snails (not even the right phylum, much less the right order).

According to Andries (1912),

Spix discovered the larva of *Microdon* near Ammerland on Lake Starenberg, in old stumps of oaks and spruces that were still rooted in the ground— and always in the company of *Formica herculanea* and *Formica rufa*. According to his own words (the larva) appeared to him at first sight like a webwork of spiders, or a footless insect larva, finally even as a turtle-like little animal. 'To the same extent that the deception disappears upon closer examination,' he continues, 'it increases the astonishment concerning its peculiar form, and the conviction gains increasingly the upper hand with the observation how the larva can creep, almost imperceptibly, on the footless, naked belly, and manages to explore nearby objects by sudden contractions and expansions of the fleshy tentacles, that this peculiar little animal does not belong among the insects which are equipped with feet and jointed feel-horns, but rather belongs in the class of the snails.' He (Spix) then expresses his delight to have found a new genus, such a beautiful addition to the snail fauna of his own fatherland.

[The translation of Andries, by the way, was graciously provided to me by Dr. Rainer Zangerl, a very fine man who is noted not only for being one of the rare individuals who actually under-stood Willi Hennig's famous book *Phylogenetic Systematics* but for actually translating it into English, so new generations of system-atists could argue about it in yet another language.]

Debate on the proper placement of *Microdon* raged on. Schlotthauber (1839) actually figured out that *Parmula cocciformis*, described as a scale insect, as well as the Fatherland's newest snail were actually the larval stages of *Microdon*. He even delivered a detailed paper to the Naturalists Congress in Pyrmont with the rousing title, "Über die Identität der Fliegenmaden von *Microdon mutabilis* Meig. mit den vermeintlichen Landschnecken *Scutelligera* (Spix) und *Parmula* (v. Heyden) sowie morphologische, anatomische und physiologische Beschriebung und Abbildung ihrer Verwandlungsphasen und ausführliche Naturgeschichte derselben. Zur Kenntnis der Organisation, der Entwicklungs- und Lebensweise aller zweiflügeligen Insekten überhaupt." His exhaustive and excruciatingly detailed study pretty much demolished the snail theory but unfortunately he never got around to publishing it, the title, perhaps, having exhausted all of his creativity. It really wasn't until 1899 that Hecht more or less came to the decision that, however snaillike on the outside, *Microdon* remained an insect at heart (or dorsal aorta).

So I have a lot of empathy with *Microdon*—here at the University of Illinois our department is more than a little like a small strange object, unfamiliar to those looking in from above, surrounded by hordes of vicious angry biting creatures bent on driving it from their ranks. I suppose my viewpoint is somewhat colored by the fact that the Illinois Board of Higher Education once targeted the undergraduate entomology curriculum for elimination (along with biophysics and astronomy) for being too "specialized." It's hard to know how to counter that kind of argument, given that there are about a million insects (and at least as many stars, as far as the Astronomy Department goes). Here in the College of Liberal Arts and Sciences there are more than 60 full-time faculty members in the English Department and there aren't even as many English words as there are insect species. There are at least as many faculty in British literature alone as there are in our entire department—not even good old American literature, but British literature. Shakespeare wrote about wars and epic battles but he never influenced the outcome of any, as did insect vectors of typhus, malaria, plague, and other wartime scourges; Wordsworth wrote lovely poems about daffodils but he never pollinated one. There's definitely a perception problem here and I just don't know what to do about it.

BUZZWORDS

I suppose I could take inspiration from *Microdon*—the maggots manage to remain untroubled in ant nests by producing analogues of brood pheromones, signal chemicals that   induce the ants to care for them and transport them lovingly throughout the colony while they happily consume their fill of ant grubs. I'll let you know if I can identify anything that works the same way on deans.

Ah! Humbug!

Anyone who has ever had occasion to grade student insect collections undoubtedly has come across composite specimens—bits and pieces of various insects painstakingly pieced together from many branches of the insect phylogenetic tree. My first encounter took place early in my entomological career—I was a teaching assistant in Entomology 212 at Cornell University. As teaching assistants were expected to do, I was grading collections when I spotted it: a single specimen that was clearly not of natural origin. Keying it out would have been a challenge but for the fact that the student thoughtfully saved us the trouble by clearly labeling it *Humbug*. As I recall, we gave him full credit for the identification.

From an entomological perspective, the term "humbug" is a bit of a disappointment. According to the Oxford English Dictionary (OED), the word has several meanings, but none of the meanings appears in any way related to insects. There's the familiar meaning of humbug as "a hoax; a jesting or befooling trick," or "a thing which is not really what it pretends to be." But a humbug is also "a kind of sweetmeat" (specifically, peppermint-flavored toffee lumps) and "a nippers for grasping the cartilage of the nose. Used with bulls and other refractory bovines." Reading about refractory bovines and toffee lumps under the heading "humbug" made me wonder whether the folks at the OED were pulling a little "jesting or befooling trick" of their own.

The etymology of the term is perhaps even less satisfying to an entomologist than the definition; according to the OED, "humbug" is a

> slang or cant word which came into vogue c 1750 (an earlier date has been given in several Dictionaries on the ground of the occurrence of the word in the title of Fred Killigrew's *Universal Jester*, which the *Slang Dictionary* dates 'about 1735–1740.' But the earliest ed. of that work is dated by Lowndes 1754). . . . Many guesses at the possible derivation of humbug have been made; but as with other and more recent words of similar introduction, the facts as to its origin appear to have been lost, even before the word became common enough to excite attention.

In other words, nobody knows why, or even whether, there is a "bug" in "humbug."

Which is not to say that arthropod humbugs do not exist. The arthropod humbug actually antedates the apparent origin of the word. No less an authority than the great Carolus Linnaeus himself, the man who named some two thousand insect species and devised the system of nomenclature in use today that is named in his honor, was taken in by a fake. In distinguishing between his new species, *Papilio ecclipsis*, and the well-known European brimstone butterfly, *P. rhamni* (*Gonepteryx rhamni*), in the twelfth edition of *Systema Naturae*, Linnaeus called attention to the distinctive black wing patches and crescent-shaped blue mark on the hindwing of the former—which, unbeknownst to Linnaeus, were painted on. This fabulous fake, along with many others, is chronicled to good effect by Peter Dance in his remarkable book

Animal Fakes and Frauds. Dance was moved to remark that the relative paucity of insect examples of humbuggery is likely due to the fact that "such fakes could only have been made to dupe a relatively small number of entomologists ... and others who derive pleasure from collecting and studying lowly creatures." He apparently never considered the equally plausible explanation that insect humbugs are few because we entomologists are simply too astute to dupe.

The earliest example of an insect fake that he recounts is found in Maria Sibylle Merian's *Metamorphosis Insectorum Surinamensium*; the 49th plate of the second edition of the book conspicuously featured an insect with the body of a cicada (*Diceroprocta tibicen*) and the head of a lanternfly (*Fulgora laternaria*). How it happened to be depicted is a mystery, but it's likely that Merian, who otherwise gave no indication in her works of having a wry sense of humor, was the butt, rather than the engineer, of the joke.

The actual word "humbug" does appear early on in an arthropodan context. J. C. Loudon's *Magazine of Natural History*, was founded with the aim of "promoting a taste for Natural History among general readers," which, in 1828, also meant protecting the public from unscrupulous charlatans anxious to capitalize on the public's fascination with natural history. The first volume, for example, featured an article on "the tests by which a real mermaid may be discovered." A letter appeared in the second volume, in 1829, signed M.C.G., describing a strange and wonderful Tarantula Sea Spider captured in a fisherman's net in the vicinity of Margate. This creature

> has eight legs, which are not jointed; and ... but two eyes which, when alive, were green, and are placed on the back of the thorax. It has no head, and is destitute of palpi. The mouth is beneath the abdomen, and inside of

it is a spiral tongue nearly half a yard long, the extremity of which is armed with a pair of forceps. The spinner is very large, out of which the exhibitor took a web, but unluckily had thrown it away.... The colour of the insect is that of a pickled tongue, which, probably, may be accounted for by the pickle that had been used to preserve it....You may form some idea of its size when I add, it weighed 5 1/4 lbs. Many wonderful stories are told of it when alive; such as it ran with the velocity of a race horse, and changed colour every instant."

Its owner, Mr. Murray of Hastings, planned to exhibit it later in the year, pickled tongue probably planted firmly in cheek. Unfortunately for Mr. Murray's business, this article was spotted by the ever-vigilant reader V., who had earlier exposed the seven-inch bison exhibited by Murray as a fraud (1829, *Mag. Nat. Hist.* 2: 218-219). About the Tarantula Sea Spider, V. ascerbically stated, "Had you inserted my article on the Pygmy bison four months ago . . . you might have saved many individuals the mortification of being humbugged by another attempt of the rare individual to appropriate some of their cash to his own use by such unfair means as the exhibition of his Tarantula or Sea Spider."

Humbuggery is a lot harder to get away with than it used to be, thanks to modern methods of analysis; today the practice is engaged in more for amusement than for profit. One of the most venerable twentieth century examples of the art of insect fakes is the big bug post card. These so-called tall tale post cards were the special effects wonders of their day, with watermelons the size of boxcars or ears of corn as tall as radio towers—inevitably with a caption along the lines of "The kind of corn we grow in Oregon, Missouri," or "How we do things in Omaha, Nebraska" or "The size we grow them at Osage, Minnesota." Archer King of Table Rock, Nebraska, was a real pioneer in producing innovative big

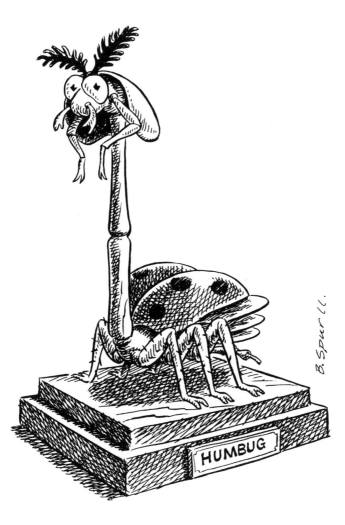

HUMBUG

bug postcards. Whereas most purveyors of gag postcards contented themselves with oversized produce, King went in for giant rabbits, giant fish, giant hogs (occasionally depicted eating giant corn), and other more challenging zoological subjects; he was, for example, the first and perhaps only post card producer to feature a giant cicada.

Undisputed master of the big bug post card, however, was one F. D. Conard of Garden City, Kansas. A relative latecomer to the field of big thing post cards, he began his business in 1935. His was a big business in every sense of the word—in his first year, he sold 60,000 postcards and two years later his annual sales exceeded 350,000 postcards. A true artiste, he dealt almost exclusively with giant grasshoppers—pulling plows ("The old grey mare she ain't what she used to be"), climbing oil rigs ("The Inspector"), crossing bridges ("Hopper has the right of way"), sporting saddle and bridle ("Ride 'em cowboy"), and even being interviewed on the radio ("A hopper tells a whopper via radio").

The entrepreneurial spirit of Conard lives on in Don Moffet, who can now rightly be considered the reigning king of big bug post cards.   Several years ago, his company, Charm Kraft of West Lake Village, California, bought out John Hinde Curteich, Inc., the original producers of post cards decades ago. Along with Curteich came its archives, as well as a postcard museum in Wauconda, Illinois, and Moffet was inspired to gear up production of a whole new line of big thing postcards—including big bug cards featuring a 25-foot-long dragonfly casting an enormous shadow across a lake ("they have been known to pick up animals and small children"), a 12-foot-long cockroach obscuring a motel sign ("these specimens really take the cake!"), and, as proof that a good joke is timeless, the ever-popular 400-pound grasshopper ("rare sport of grasshopper shooting can be dangerous and exciting"). The images are slicker than they've ever been before, Moffet and his photographer son Buddy having been assisted by computers in creating them, and the cards are selling well. All are emblazoned on the front with the slogan "Bigger and Better in America" in big bold letters. Not so obvious, in very small type along the bottom at the back of the card, is written, "Printed in Ireland." It looks like the humbug is alive and well and, like so many other strange things, living in California.

Grumpy old entomologists

As most entomologists will aver, the benefits of taking up entomology as a profession are neither readily apparent nor easily articulated. It's not as if a Ph.D. degree in entomology is a ticket to instant fame, fortune, and success in love. A disturbingly large proportion of the general public isn't even aware of what an entomologist is, and, of those who actually do know what an entomologist is, a disturbingly larger proportion doesn't know exactly how to spell the word. But, ironically, one of the greatest benefits of this particular career choice is something that I would bet most entomologists themselves are unaware of. Whatever the indignities they must endure because of their profession, at least entomologists should have the satisfaction of knowing that they are remarkably durable.

I first became aware of the amazing longevity of entomologists in the usual roundabout way I become aware of most interesting things—I was looking for something else. In this case, I was in pursuit of an article on cockroach feeding preferences written by Phil Rau in 1945 and published in the journal *Entomological News*. While paging through the journal, I stumbled onto another interesting paper, H. B. Weiss's "How Long Do Entomologists Live?" in the same issue. Dr. Weiss had evidently made a detailed analysis of another paper published that same year, M.M. Carpenter's 116-page "Bibliography of Biographies of Entomologists." This compilation included birth and death dates for some 2,187 entomologists active between 372 BC and 1920. According to Weiss' analysis, the average age at death of this group was 65.48 years. Remarkably, breaking down the group by century and calculating life expectancies didn't really change the average much—an entomologist was just as likely to reach the ripe old age of 65.48 in 1605 (when, for example, systematist Ulysse Aldrovandi died, at age 83) as in 1905 (when, for example, aphidologist George Bowdler Buckton died, at age 88).

Thus, as Weiss (1945) pointed out, entomologists have been outliving their contemporaries for centuries, by staggering margins. In Breslau, Germany, in 1685 the life expectancy of an entomologist was approximately twice that of a typical male resident, which at the time clocked in at about 34 years. A century later, in England, the average life expectancy of a typical male was only 40, fully 25 years less than that of a contemporary entomologist, and in the U.S. in 1910 the average life expectancy was up to 50, 15 years less. By the time Weiss wrote his paper, the life expectancy in the U.S. for males had risen to 62.94, 2.54 years less than for the typical entomologist.

Weiss was amazingly matter-of-fact about his findings—he succinctly concluded his paper by suggesting that, because heredity is largely responsible for lifespan, "most of the credit for living long lives should go to the parents of entomologists." I, on the other hand, was extremely troubled by the pattern that he had documented. What I saw was an alarming erosion of the differential in lifespan between entomologists and the general populace. It was disturbing to me that, since the life expectancy of entomologists had been twice that of their contemporaries in the seventeenth century, it wasn't also twice the life expectancy of contemporaries (on the order of 125.88) in 1940, when Weiss wrote his paper. Given the seemingly inescapable fact that the one relative advantage of being an entomologist was in danger of fading away, I searched the literature to see if the trend had continued in the intervening years since Weiss' paper appeared.

Despite some effort on my part, I was successful in turning up only a single paper on the subject of entomologists' longevity, published in 1976 in the little-known journal, *Insect World Digest*. In the paper, Messersmith reported finding a table in a book called *Man*, by R. J. Harrison and W. Montagna and published in 1969, reporting longevity of "eminent men" according to profession. This table actually included entomologists among its eminent men; also in these ranks were philosophers, historical novelists, state governors, authors of church hymns, and composers of both choral and chamber music. According to Harrison and Montagna, entomologists in 1969 had a life expectancy of 70.89, exceeding that of any other profession except for members of U.S. Presidents' cabinets. Entomologists in fact outlived their fellow botanists (68.36 years), chemists (69.24 years), geologists (69.79 years), and mathematicians (66.62 years). For completeness' sake, it is perhaps worth pointing out that the shortest life expectancy reported was for hereditary European sovereigns, who could expect to live a mere 49.14 years.

This book, *Man*, by the way, is a little strange. It seems to be a textbook of some sort, written by a team consisting of a male marine biologist and a male primatologist. If you wonder why it struck me as peculiar, it's that I find it hard to understand why a book with the title *Man* has such an extraordinary number of photographs of naked female breasts—16, by my count (photographs, not breasts).

My efforts to turn up another, more recent, survey were unsuccessful. To get an approximate idea of the current status of our advantage over our peers, I turned to the pages of the *American Entomologist*, the official publication of the Entomological Society of America. A review of obituaries in the pages of this journal was anything but reassuring. Of the 172 men and one woman whose obituaries were published in this journal between 1983 and 1996, the average lifespan was 72.5—well below 75.7, the years of life expected at birth for men and women of all races born in 1994.

I suppose it shouldn't be too surprising that the rest of the world is catching up with us. Truth be told, I can't figure out why we had it so good for so long. Particularly in recent centuries, being an entomologist meant either spending an inordinate amount of time at a desk, hunched over tiny specimens and impaling them with pins while writing out minute labels in a crabbed hand, or in an agricultural field, soaking up or breathing in noxious organic compounds designed to short-circuit the nervous system or accumulate in body fat or breast milk. Neither set of activities seems especially conducive to long life. But the study of occupational health and longevity is often full of contradictions.

These contradictions were well known to one of the earliest practitioners of occupational health and medicine, William Thackrah (1795-1833), a British surgeon and apothecary who in 1832 authored one of the first definitive studies of the association between profession and disease, a book titled *The effects of arts, trades, and professions, and of civic states and habits of living, on health and longevity*. As might be expected, the professions he studied differed from what might be studied today. Among those he catalogued in the garment industry were scribblers and carders of wool, slubbers of cloth, spinners, weavers, raisers of cloth, croppers, burlers, frizers, cloth-drawers, and blanket makers, most of which rarely, if ever, turn up in the "Help Wanted" section of the newspaper these days. I found no heading for "entomologists" per se but on reading through the book found the category into which I suppose he was most likely to have placed them, had he encountered them (p. 180), "the last class of society—persons who live in a confined atmosphere, maintain one position most of the day, take little exercise, and are frequently under the excitement of ambition. This class includes individuals from the several professions, as well as the men devoted to science. . . ."

Thackrah provides no insight as to why entomologists were long-lived; if anything, his litany of the ills facing scholars would lead one to the opposite prediction. According to his observations,

> The position of the student is obviously bad. Leaning forward, he keeps most of the muscles wholly inactive, breathes imperfectly, and often irregularly, and takes a full inspiration only when he sighs. He generally lives, too, in an impure atmosphere, and neglects the common means of relief. The circulation is enfeebled; the feet become cold. The appetite. . . whether moderate or excessive . . . is greater than the power of digestion; for the application of mind too great or too long, absorbs that nervous energy, which digestion requires. The stomach becomes foul, the secretion of bile is impaired or vitiated, the bowels are sluggish, and constipation, with its attendant evils, progressively succeeds. The brain becomes disturbed. . . . A highly excitable state of the nervous system is not infrequently produced. Irritability of temper, vain fear and anxiety about trifles, mark, in common life and ordinary circumstances, the character of men. . . . Chronic Inflammation of the membranes of the brain, ramollisement of its substance, or other organic change, becomes established; and the man dies, becomes epileptic or insane, or falls into that

imbecility of mind, which renders him an object of pity to the world, and of deep afflictions to his connexions.

So I guess I've found my answer. I had assumed that long life meant long, healthy life; evidently, although we collectively may live longer, we entomologists live out our earthly span as irritable, constipated burdens to our relatives. Maybe that is an advantage to our profession after all—if we can't have fame, fortune, and good health, at least we can provide annoyance to those who do have them.

By the way, the central premise of this essay—that entomologists have until recently enjoyed a longer lifespan than most other people—is based on a distortion of demography in that average lifespans should be calculated from birth. Because it is impossible to identify at the moment of birth those individuals destined to become entomologists, average lifespan calculations are compromised; unlike the apparently short-lived European sovereigns, entomologists are made, not born. In my defense, I can say that I was only extrapolating the logic used by Harry Weiss in the 1945 paper that inspired this essay. Based on my reading of several other articles he had written, it is my guess that Dr. Weiss, too, had his tongue at least partially implanted in his cheek when he wrote his paper. Incidentally, playing fast and loose with life tables apparently did Dr. Weiss no harm—he lived to be 91 years old. Not so the author of the paper on cockroaches I was originally looking for, Phil Rau. Despite his interest in longevity—he even authored a paper in *Annals of the Entomological Society of America* (38:503-504, 1945), titled, "Longevity as a factor in psychic evolution," in which he suggested that "the high mental qualities in the animal world are the result of long life"—Rau died in 1948, at the unentomological age of 53.

Images of entomologists— moving and otherwise

Probably everyone agrees that stereotyping people is bad, but there are some people that can be stereotyped with aplomb without fear of societal disapprobation. Entomologists are among those people. I've been photographed on several occasions for a variety of types of publications—newspapers, magazines, and the like—and it seems that, every time, photographers ask me to pose in one of three ways: seated in front of a microscope; with an insect, usually a cockroach, on my face; or with an insect net clutched in my hand. Posing with a microscope is all right, I guess, but it's been done to death, not just with entomologists but with life scientists of all descriptions. I categorically refuse to put any kind of insect on my face; as I explain to the photographers, there's no earthly reason that I can think of for any kind of entomologist to walk around with arthropods on his or her face, the human follicle mite, *Demodex follicularum* (which lives in the follicles of human facial hair) notwithstanding. For the record, and for any photographers who might be reading this article, I also categori-cally refuse to pose with deely-boppers, wings, or any other prop designed to make me look like an insect. I don't mind posing with a net, but, again, most of the time these photographers are shooting pictures indoors and I explain to them that there are very few occasions upon which I must use a net while in my laboratory. Also, for the record, I won't pose wearing a pith helmet or a safari jacket, either, articles of apparel that most photographers seem to feel hang in every entomologist's closet.

I give photographers a hard time basically because I have no desire to look ridiculous. This aversion of mine takes them by surprise because it's their perception that entomologists ought to look ridiculous. It's not hard to figure out where this perception comes from—entomologists have had an image problem ever since the discipline came into its own. Examples from popular literature are rife. Among the earliest references I could find to an entomologist in a work of fiction comes from a story written in 1895 by H.G. Wells, called "The Moth." This story is an account of a feud between "the celebrated Hapley, the Hapley of Periplaneta Hapliia, Hapley the entomologist" and one Professor Pawkins. The feud began

> years and years ago with a revision of Microlepidoptera (whatever these may be) by Pawkins, in which he extinguished a new species created by Hapley. Hapley, who was always quarrelsome, replied with a stinging impeachment of the entire classification of Pawkins. Pawkins in his 'Rejoinder' suggested that Hapley's microscope was as defective as his powers of observation, and called him an 'irresponsible meddler' . . . Hapley in his retort, spoke of 'blundering collectors,' and described . . . Pawkin's revision as a 'miracle of ineptitude.'

What eventually happens is that Pawkins ("a man of dull presence, prosy of speech, in shape not unlike a water-barrel . . . and suspected of jobbing museum appointments") dies before replying to Hapley's stinging critique of his work on the 'meso-blast' of the Death's Head Moth. His sudden demise leaves Hapley (of "disordered black hair [and] queer dark eye flashing") without a purpose in life. To make a 10-page story short, Hapley begins to hallucinate, imagining himself to be pursued by a strange moth, invisible to others, that bears an uncanny resemblance to the deceased Pawkins. The story ends with Hapley "spending the remainder of his days in a padded room, worried by a moth that no one else can see. . . . "

Thus, things didn't start off very auspiciously for entomologists in literature, and, unfortunately, things haven't improved much since then. Even when entomologists are sympathetic characters, they invariably look peculiar or act eccentric. Witness Shelly Hubbard, the entomologist in *Swarm*, a 1974 Arthur Herzog novel about killer bees:

> . . . Hubbard resembled his own zoological specialty, the beetle. At fifty-two he was short and massively broad, with a bulging chest. He had almost no neck at all, and his round head seemed to pivot directly on his sloping shoulders. Two fringes of black hair stood up on the sides of his bald dome like antennae. Habitually, and in line with his coleoptera character, Hubbard rubbed his hands together with a rustling sound or created sucking noises by making a vacuum between his palms.

Interestingly, there is another scientist in the story, an environmental biologist, described most emphatically as "not an entomologist, much less a bee man." In contrast with his colleague, he "was thirty-five, six feet tall, thin, with a straight nose, blue eyes, angular cheeks, an affable but controlled mouth and a military set to his shoulders. He had the sort of face women reacted to. . . . " Women probably reacted to Hubbard, too, most likely by screaming and running away in terror.

Then, there's Noble Pilcher, in Thomas Harris' 1988 *Silence of the Lambs*. He's the Smithsonian entomologist who assists Special

How entomologists see themselves

Agent Clarice Starling in tracking down a serial killer who leaves insects at crime scenes—"Pilcher had a long friendly face, but his black eyes were a little witchy and too close together, and one of them had a slight cast that made it catch the light independently." Agent Starling first encounters Pilcher in his office as he and fellow entomologist Albert Roden are absorbed in a game involving a rhinoceros beetle and a chessboard and arguing heatedly over the rules.

When "Silence of the Lambs" was made into a movie, it was only natural to include the character of Noble Pilcher for comic relief. Hollywood has been even harder on entomologists than has the literary world, particularly when they're ancillary characters. While it's true that positive depictions can be found of entomologists who are presentable and reasonably normal in appearance and who haven't unleashed some kind of hexapod plague upon humanity, it's also true that such films are few and far between. It's far easier to find cinematic examples of entomologists with thick glasses and no sense of style in terms of apparel (and just because I happen to wear thick glasses and lack any sense of style in terms of apparel doesn't mean I resent the stereotype any less). In "Fierce Creatures," a John Cleese film about a zoo in England, Adrian Malone, the "keeper of the Marwood insect house, was renowned for his loquacity. Rarely has a human being on the planet Earth been quite so verbally unchallenged. To Bugsy—a soubriquet which he had once been unwise enough to say he detested and which as a result had become the name by which he was universally known . . . life was one long soliloquy. And that soliloquy was primarily on the subject of insects." Suffice it to say, nobody likes him. Nor does anybody appear to like Dr. T.C. Romulus, the entomologist in "BioDome," who sports not only the inevitable thick glasses and pith helmet but, inexplicably, a flyfishing vest as well. (Note to readers: the fact that there is an entomologist in the film is the ONLY REASON I actually bought a video with Pauly Shore in it.)

Whatever else you might think of it, television may actually be the salvation of the entomologist. Although they're not commonly encountered on the small screen, when entomologists do appear, they're surprisingly sympathetic. Among the greatest achievements in advancing the image of the entomologist with the public was the appearance of Dr. Maxsy Nolan, of the Department of Entomology at the University of Georgia, on the television show, "Space Ghost Coast-2-Coast." For those who sleep regular hours and thus may not be familiar with the Cartoon Network's late-night offerings, SG–C2C is a part animated, part live-action extraterrestrial talk show featuring host Space Ghost, a superhero who first appeared in Hanna Barbera cartoons in the 1960s, and his sidekick/keyboardist Zorak, a giant alien mantis. Space Ghost interviews real-life celebrities on a television monitor while Zorak mutters generally disparaging remarks and threatens to destroy things. Dr. Nolan was a guest on episode 41, titled "Zorak," which was kind of a "This is Your Life" retrospective in honor of Zorak. Dr. Nolan and another guest, an exterminator, offered insights into the lives of mantids. I asked Dr. Nolan by e-mail how he enjoyed his stint and he admitted that he "had a whale of a time with the show," dealing, among other things, "with a nine foot mantis hovering over me the entire time of the shooting (about 4 hours)." What makes this appearance such a landmark in the stereotype-busting is the fact that Dr. Nolan has joined an extraordinary elite, individuals that define popular culture. Other guests of SG C2C include such luminaries as actor Charlton Heston, psychologist Dr.

How entomologists see themselves

Joyce Brothers, rapper Ice-T, cartoonist Matt Groening, astronaut Buzz Aldrin, and parodist (and personal favorite) "Weird Al" Yankovic.

In terms of prime-time achievements in stereotype-busting, however, recognition must go to the Fox Network program "The X-Files," a show that depicts the activities of a division of the Federal Bureau of Invesigation devoted to investigating inexplicable and potentially paranormal phenomena. Of particular significance was the episode that aired originally on January 5, 1996, titled, "War of the Coprophages." In this episode, Agent Fox Mulder is called in to investigate a mysterious series of cockroach-related deaths; in time, Mulder becomes convinced that these are no ordinary cockroaches and may, in fact, be extraterrestrial in origin. As he pursues his investigation, he eventually teams up with a U.S.D.A. entomologist he encounters after breaking into her laboratory (where she's been investigating, among other things, the electrical properties of cuticle and the effects of light, temperature, humidity, and food availability on behavior). Here's how the novelization of the episode (Martin 1997) describes the encounter:

> Standing in the doorway was the best-looking woman Mulder had seen in a long time. Her eyes were bright against her dark hair. Her flannel shirt, safari shorts, and hiking boots looked surprisingly attractive. But the look on her face told Mulder that she was not nearly as impressed with him. In fact, she looked downright angry.
>
> 'What do you think you're doing here?' She demanded. . . . 'This is government property. And you are trespassing.'
>
> 'I'm a federal agent,' Mulder said.
>
> The woman's eyes didn't soften. 'So am I.'

BUZZWORDS

Mulder put his phone back in his pocket. He flashed his badge.

'Agent Mulder—FBI,' he said.

'Dr. Berenbaum,' the woman said. 'U.S. Department of Agriculture Research Service.'

'Dr. Berenbaum,' Mulder said. 'I need to ask you a few questions.'

'For instance?' the woman said.

'What's a woman like you doing in a place like this?'

Dr. Bambi Berenbaum then proceeds to lecture Mulder on the habits of cockroaches and ably assists him throughout the episode with her entomological expertise.

When I first became aware of this episode, I naturally took a considerable interest in it. Two questions came to mind. First, in the *TV Guide* listing for the episode, I couldn't help noticing that the fictional entomologist's last name was identical to mine. It was hard to believe that this might be by chance because not even all of my relatives spell "Berenbaum" exactly that way. More important, was the casting of gorgeous actress Bobbie Phillips in the part of Dr. Berenbaum a desperate ploy for ratings on the part of a casting director, or was it the actual intent of the scriptwriter to make the entomologist an attractive character? There was only one way to find the answers to these questions—to go to the source, scriptwriter Darin Morgan.

It took me almost two years to work up the nerve (among other things, Darin Morgan is revered by "X-files" fans and is something of a celebrity as a result), but I finally spoke with the man and found him charming, gracious, and extremely personable. He told me that he had consulted some of my books in preparing the script and thus felt "Berenbaum" would be an appropriate name for an entomologist. In response to my question about Dr. Berenbaum's appearance, he replied that he had indeed intended to depict her as a "luscious babe. . . . I needed a rival [for female Agent Scully] and it helps if she's really good-looking."

Kudos, then, to Darin Morgan for actually conducting some

Stereotype specimen

BUZZWORDS

research before putting pen to paper and for rising above stereotypes. Thanks to him, we've come a long way—from "water barrel" to "luscious babe," to be precise. As for the next photographer who asks me to put on a pith helmet or to kiss a cockroach, I think I'll just show him a copy of my Bambi Berenbaum X-Files Collector Card and ask him to reconsider that request.

(Water) penny for
your thoughts?

As an entomologist, I have often wondered whether my parents, protestations to the contrary, may not have been the teensiest bit disappointed with my career choice. Admittedly, they have always been embarrassingly proud of my entomological accomplishments, such as they are. When I wrote a book, a collection of what I hoped would be regarded as humorous essays about insects titled, *99 Gnats, Nits, and Nibblers*, they made sure that every single one of my relatives, no matter how distant, received his or her own copy. This includes family members whose native language isn't even English and babies who haven't yet learned how to read. I'm sure the University of Illinois Press sales staff can't figure out why so many book orders keep coming in from the state of New Jersey; I think my parents alone are responsible for the fact that the book went into a second printing.

Nonetheless, when the journal *Science* ran an article in 1991 about "Career Trends in the 90s," my father, a polymer chemist with a major chemical company, called me up to point out that, according to the article, entomologists are paid less than scientists at equivalent rank in any of the 17 life science disciplines surveyed. I had missed the article the first time through the journal (probably too busy sailing my yacht or exercising my polo ponies), but, intrigued, I went back to find my copy and read the article. My father had somewhat overstated the case—it was true that full professors of entomology made less than anyone else, but assistant professors of entomology on average earned more than assistant professors of biology, botany, marine biology, and zoology. But it was undeniably apparent, in full color graphics, that my chosen profession is hardly a lucrative one.

Although the details were interesting to see, the fact itself was hardly a revelation to me. I don't think anyone goes into entomology to earn heaps of money and to win the respect and admiration of one's peers. And this fact is hardly a product of contemporary crass commercialism. While rummaging around the entomology department archives, I came across a broadside that had been in the reprint collection of W. P. Hayes, department head from 1945 to 1953. On the back it was stamped "FEB 10 1920": on the cover was written:

AMERICAN ENTOMOLOGISTS
and those Employing Entomologists
will find in this Paper a VALUABLE MESSAGE
American Entomology:
Its Present and Future Status as a Profession

This paper is published and distributed by a group of younger Entomologists who are concerned with the advancement of the science they love. It is their fondest hope that Entomology will shortly be placed on such an improved basis that they will be able to devote their uninterrupted thought and effort to the subject without endangering the welfare of their homes.

The tract was basically a lament about the meager salaries paid to professional entomologists. Such salaries meant that young men who wished to be entomologists were forced to find outside work, or even to abandon the profession altogether, to maintain hearth and home. Those choosing to remain in the profession were obligated to "beg at corner cigar stores for boxes in which to store his specimens." The anonymous author of the tract implores "those superior in power and influence" to remedy the situation— "by entering the field of Entomology one should by no means infer that he must become a vagabond." That the author was not optimistic is evidenced by his concluding paragraph—

> . . . then the Entomologist may walk boldly down the principal street of the city and look his fellow citizens in the eye, rather than slinking thru the back alleys so that his ragged appearance will not be noticed. Then will the hands of fellow citizens and fellow scientists be offered in respect and with honor to the Entomologist and he will no longer be greeted with a smile of amused ridicule—then, perchance, the millenium will arrive.

Things were undoubtedly worse then than they are now, no matter how small your raise was last year—after all, the millennium has arrived and it IS possible to make a living as an entomologist these days. Of the 16 founding members of the Entomological Society of Washington, only about half were paid to be entomologists. Reverend J.G. Morris was a clergyman of the German Lutheran church in Maryland, Lawrence Johnson was a judge, E.S. Burgess was a botanist who taught high school biology, C.J. Schafhirt was a druggist, Alonzo H. Stewart was a page in the U.S. Senate, R.S. Lacy was a Washington lawyer, R.W. Shufeldt was an ornithologist, and John Murdoch was a librarian. Ten years after its founding in 1884, L. O. Howard exhorted its members not to lose a single opportunity to press the importance of . . . a donation to science in particular, and to the world at large, upon chance millionaires of their acquaintance. . . . Who knows but a clause may be found in the will of some one of the men who are already active members of our Society, which will put us upon a firm financial basis? We are not looking forward to the demise of any of our wealthy members, and hope that they may be with us for many years to come. When, however, full of years and full of honor, they prepare themselves for the inevitable end, let us hope that . . . a little slice of their accumulated riches may be left to the struggling organization upon which they have shed the lustre of their names.

To some extent, I think entomologists bring financial penury upon themselves. In general, entomologists are all too willing and eager to dispense their knowledge and practice their skills gratis. Bring your car to a mechanic and you have to pay him thirty dollars just for him to tell you what's wrong with it; bring a wilted coleus plant to your neighborhood entomologist and he or she will not only identify what's causing the problem but will provide you with enough reprints, bulletins, and other assorted reading material to last you a week, all absolutely free of charge. This practice is so ingrained that making money at entomology is almost regarded as dishonorable. Alexander Arsene Girault, for example, who worked for the United States Department of Agriculture and for the Australian government as an entomologist near the turn of the twentieth century, was particularly adamant on this point. He "felt that the use of entomology for economic purposes was a prostitution of science and learning" and wrote derogatory doggerel about his colleagues who profited from their profession (Spilman 1984). At one point, he lambasted J.F. Illingworth of Australia in a snide, sarcastic parody of a scientific description—"Shillingsworthia."

How entomologists see themselves

S. shillingsworthi, blank, vacant, inaneness perfect. Nulliebiety remarkable, visible only from certain points of view. Shadowless. An airy species whose flight cannot be followed except by the winged mind. . . . This so thin genus is consecrated to Doctor Johann Francis Illingworth, in these days remarkable for his selfless devotion to Entomology, not only sacrificing all of the comforts of life, but as well as his health and reputation to the uncompromising search for truth.

There probably remains a bit of residual Girault in all of us, accustomed as we are to being overlooked and unappreciated.

BUZZWORDS

What we lack in power, wealth, and prestige, perhaps we make up for in nobility of purpose and self-sacrifice. By the way, in case you're wondering, I'm giving away half of the proceeds from this book to the Entomological Society of America, which never paid me in the first place to write these essays. And no, I haven't had the nerve to tell that to my father yet.

Rated GP ("generally patronizing")

emale entomologists are not now, nor have they ever been, particularly numerous. There are so few of them, in fact, that it seems unlikely that anyone (with the possible exception of a male entomologist) has had contact with enough female entomologists to form any sort of opinion about them at all. And yet prejudice and hostility toward female entomologists exist.

Am I, as a female entomologist, just being paranoid, you ask? I hardly think so. All you have to do is take a look at how female entomologists are depicted in insect fear films. Admittedly, scientists in general don't come off too well in this particular genre, but at least occasionally male entomologists are positively heroic. Dr. Harold Medford, the kindly old "myrmecologist" in *Them* (1954), for example, saved Los Angeles from a swarm of giant ants. He even looked saintly: the man who played the part, Edmund Gwenn, had just a few years earlier played Santa Claus in *Miracle on 34th Street.* Handsome, young Peter Graves in *The Beginning of the End* (1954) singlehandedly saves Chicago from an atom bomb that the army had planned to drop to rid the city of a plague of thirty-foot locusts. Granted, it was his own sloppy experimentation with radiation that produced the giant locusts in the first place, but the audience forgets such details by the last reel of the film.

Even the crazed male entomologists in these films are at least well intentioned. Dr. Deemer (Leo G. Carroll), the scientist in *Tarantula* (1955) who accidentally looses a thirty-foot tarantula on an unsuspecting town, was only trying to develop a synthetic food to save millions from starvation. Peter Graves (*The Beginning of the End*) was using radiation in the first place to grow giant tomatoes, among other things, to feed the hungry masses.

Female entomologists, on the other hand, have only one thing in mind: achieving eternal youth and beauty. For years, female scientists in movies have had the peculiar conviction that insects or their various bodily fluids have pharmacological properties that can bestow beauty and longevity upon those who consume them, a conviction that is all the more peculiar given the physical appearance and breathtakingly brief life span of the vast majority of arthropods. Generally, these women are not even interested in developing beauty creams to save millions from the ravages of age—they usually have a vested personal interest in this research.

Take Janice Starlin (Susan Cabot), the subject of *Wasp Woman* (1959), for example. As founder and CEO of Janice Starlin Enterprises, a cosmetics firm that's floundering because she's showing her age and it's affecting her image, she hires the dubious Dr. Eric Zinthrop to prepare extracts of wasp royal jelly. When injected, these "enzymes" take eighteen years off her age and restore her to her former beauty. They also have the unfortunate side effects of stimulating the growth of antennae and creating a ravenous appetite for human blood, which might well prove to present problems in obtaining FDA approval (although it presents interesting possibilities to the advertising department).

How entomologists see themselves

As Dr. Elaine Frederick in *Flesh Feast* (1970), Veronica Lake conducts "rejuvenation" experiments in the basement of a mansion in Miami Beach. These experiments consist of allowing *Calliphora* blow fly maggots to feast on human flesh, clearing away dead skin cells to leave a younger, fresher face. One of the more admirable women scientist in insect fear films, she does, to her credit, use her maggots to destroy Adolf Hitler at the end of the film, in a series of plot twists that are too complex to describe here.

Even as experimental subjects, women in these films are embarrassingly shortsighted and selfish. In *She-Devil* (1959), a female patient receives an experimental drug derived from *Drosophila* serum that allows her to metamorphose at will. She uses this extraordinary power and once-in-a-lifetime gift to change from brunette to blonde. In *Invasion of the Bee Girls* (1973), the only soft-core pornographic insect fear film (to date), a female entomologist creates a society patterned (loosely) after that of the honey bee. Women in the society recruit new female members by having profoundly energetic sex with the husbands of the recruits-to-be, leading to massive coronaries; the grieving widows are then metamorphosed into bee girls. Among other things, metamorphosis entails a new hairdo (not inappropriately, a beehive).

An obsession with hair is a peculiar undercurrent throughout films of this genre. The doomed but noble Leo G. Carroll in *Tarantula* has a female laboratory assistant named "Steve." Many of the female scientists in insect fear films, by the way, have men's names—just another mechanism for introducing some levity at a dramatic moment. The leading man gets to do a double-take when he realizes that the scientist he'll be working with has two X chromosomes and is wearing a skirt. Just as developments in the laboratory are getting exciting, Steve remarks, "Science is science but a girl must get her hair done," and departs, leaving her test tubes behind without a second thought.

This strikes me as incredibly unrealistic, although admittedly my objections sound like nitpicking given that the movies I'm talking about also contain tarantulas or grasshoppers the size of mobile homes. In most other bad science fiction films, all the women have to do is stand around and look helpless while the male hero figures out what to do. Occasionally, they get to undress in front of windows into which giant apes or lizards peer. What is it that women entomologists have done to incur the will of at least four decades of filmmakers? Maybe playing around with insects is considered unfeminine. Perhaps male entomologists, feeling threatened by general perceptions that people who play around with insects are less manly, unconsciously are projecting animosity toward women who would perpetuate such stereotypes. Maybe there is a Jungian association in the collective unconscious of filmmakers between the image of predatory female insects and domineering mothers. I could speculate all day, but I really have to go now because I have an appointment to get my legs waxed. After all, science is science. . . .

How
an entomologist
sees
science

Author! Author! et al.

Ever since I became a department head, I have been a regular reader of *The Scientist*, a biweekly newspaper that is published by the Institute for Scientific Information in Philadelphia, Pennsylvania. The free subscription I receive as a consequence of being a department head is, as far as I can tell, the only perk associated with the job (and it doesn't even come from the campus). *The Scientist* is the closest thing there is to *People Magazine* for scientists—it runs feature stories on prominent personalities in science, reports on recipients of scientific awards and prizes, and keeps tabs on research trends. There are obvious differences, of course, between *People Magazine* and *The Scientist*. In *The Scientist*, for example, you find advertisements for real-time digital fluorescence analyzers instead of, say, Ben and Jerry's Ice Cream. Both periodicals publish reviews, but, while a May 1994 issue of *People* included reviews of upcoming episodes of "MacGyver" and "The New Adventures of Captain Planet," that same month *The Scientist* chose to review, among other things, "BDNF mRNA expression in the developing rat brain following kainic acid–induced seizure activity." And, in all the time I've been reading *The Scientist*, I've never once seen Oprah Winfrey's picture in it.

BUZZWORDS

What I did see in the April 4, 1994 issue was an article titled, "1993's Top Ten Papers: Superconductor report surfaces in sea of genetics." This article listed the "hottest articles in science for 1993—as determined by citation analysis." Citation analysis is basically the evaluation of bibliographies in published papers. The logic behind counting up citations is that the more people cite a particular publication in their own work, the more interest there is in the paper and the greater its impact—not an unreasonable assumption. Two things struck me about this article on Top Ten Papers as I ran down the list. First, there were no papers on the list with even marginal entomological content. Second, I couldn't help but notice that no paper had fewer than four authors, and the Number One Most Frequently Cited Paper of 1993, by Rosen et al., had 33 authors.

You won't find the complete citation, with all 33 authors' names, in *The Scientist*. For that matter, you won't find it in *Biological Abstracts*, which for years has listed only the first ten or so authors of a multi-author paper. And you probably won't find all 33 names in any textbook in which the paper is cited, because a lot of textbook publishers now are limiting citations to ten or fewer names. I don't even think you'll find all 33 names in the vitae of all of the authors. At least I think if I were, for example, 28th author, I wouldn't want to devote a full half-page of my vita to the 27 names that precede mine.

When you think about it, it's not really so remarkable that Rosen et al. became a citation classic so quickly. Even if each co-author cites it only once in a given year, that's 33 citations right there—although in fairness it must be said that Rosen et al. has been cited 54 times since its publication, so there are 66% more citations than authors. The work itself is unquestionably important—the study described in the paper provided a possible causal genetic mechanism to account for amyotrophic lateral sclerosis, a fatal degenerative disease of the nervous system. But it's not beyond the realm of possibility that multiple authorship can give a paper a boost in the citation department. In fact, for *The Scientist*'s top ten list for 1993, there is a product-moment correlation of 0.62 between author number and citation number, which is marginally significant at $p = 0.056$. So there's a good chance that a paper with a lot of authors will be cited a lot.

Which then led me to wonder which 1993 entomology paper, on statistical probabilities alone, should be top contender for greatest number of citations. In my search for papers with a large number of authors, I came across a few impressive ones outside of entomology, the most notable being the 42-author study of "Phylogenetics of seed plants: an analysis of nucleotide sequences from the plastid gene *rbcL*" by Chase et al. in the *Annals of the Missouri Botanical Garden*. Then there was the paper by Partsch et al. in *Hormone Research* ("Comparison of complete and incomplete suppression of pituitary-gonadal activity in girls with central precocious puberty—influence on growth and final height") with 41 authors. But in terms of papers with entomological content, I was far less successful.

A perfunctory stroll through the unbound journals at the UIUC Biology Library produced only one potential contender for multiple entomological authorship laurels—"Development of recombinant viral insecticides by expression of an insect-specific toxin and insect-specific enzyme in nuclear polyhedrosis viruses" by B. D. Hammock and 12 co-authors. However impressive 13 authors may be among entomological papers, it barely qualifies as a multi-author paper compared with the competition.

The year 1994, however, may have been a bad one for collaborative entomology. Historically, there have been entomological papers with many more authors than 13. Back in 1981, Dr. William Horsfall, my colleague here at the University of Illinois, showed me a paper he had found documenting mosquito distributions in the then-Soviet Union with more than 50 authors. I never did get a copy from him, and I couldn't find it in the library. But it's still out there somewhere, I'm sure, even if the Soviet Union isn't around anymore. Dr. Alan Renwick, of the Boyce Thompson Institute, however, could find (and did share) his copy of Hurter et al., 1987, "Oviposition deterring pheromone in *Rhagoletis cerasi* L.: Purification and determination of chemical constitution," in the journal *Experientia*, with 15 authors. And, according to Dick Beeman of the USDA Grain Research Laboratory in Manhattan, Kansas, fifteen also appears to have been an all-time high for the remarkable E.B. Lillehoj, a research chemist from the USDA Northern Regional Research Center in Peoria, Illinois. While working on aflatoxin contamination of corn and its relationship to corn-infesting insects, Lillehoj was, between 1978 and 1980, senior author of one paper with 15 authors, one paper with 14 authors, and, most remarkably, one paper with 11 authors. The last paper is most remarkable in that the eleven authors worked in 11 different institutions in 11 different states. It's important to point out, to place this achievement in the proper context, that this feat was accomplished long before e-mail was in widespread use.

Multiple authorship wasn't always the rule in entomology. In fact, in the first issue of the *Jounral of Economic Entomology*, published in 1906, only 6 of the 67 papers had more than a single author (and of those six, none had more than two authors. In fact, 3 of the 6 were co-authored by the same individual, one Wilmon

Newell, a man clearly ahead of his time). In contrast, the December 1993 issue of the same journal contained 34 papers, only 3 of which were by a single author. One paper had seven authors. Even the *book review* had two authors.

There's a lesson here, apparently—to attract notice today, it doesn't hurt to collaborate. In fact, I would like to encourage entomologists to take a run for the top of the hottest paper list. I should warn you, though, that the competition may be getting tougher, particularly from the medical community. There's a paper in a 1994 issue of *New England Journal of Medicine*, "Effect of vitamin E and beta carotene on the incidence of lung cancer and other cancers in male smokers," with 52 authors, and another paper, on clinical trials of a clot-buster drug used in coronary angioplasty, with at least 175 authors (I stopped counting after a while). This is a level of collaboration that may prove unbeatable by entomologists. I don't think even Lillehoj could rise to this challenge.

I'm okay—
are you O.K.?

Several years ago, I attended a symposium in West Lafayette, Indiana, on cytochrome P450s. P450s make up a multigene family of proteins that in insects are involved in pheromone synthesis and metabolism of xenobiotics, among other things. I was scheduled to give the fifth talk of the day, a slot for which, on that occasion, I was particularly grateful. For one thing, it wasn't the slot immediately after lunch when people fall asleep; it also wasn't the last talk of day when the audience would include only those driving home with me in the same vehicle. Most importantly, on this occasion I was relieved to be giving the fifth talk of the day because, despite having published papers on insect P450s for more than a decade, I wasn't exactly sure how to pronounce them.

Back when I first got interested in these enzymes, they were called MFOs—mixed function oxidases—because they catalyze a variety of oxidative reactions. Unbeknownst to me at the time, the powers that be had decided the name was not descriptive enough and these enzymes should from that point on be known as PSMOs, or polysubstrate monooxygenases, because they attach a single oxygen atom to many different substrates. By the early

1990s, an elaborate system of standardized nomenclature was advanced whereby these enzymes were classified into families and subfamilies, based on degree of protein sequence similarity. Now, P450s are referred to by the acronym CYP, for *CY*tochrome *P*-450. This acronym is followed by a number, which designates a gene family (>40% sequence identity); the number is followed by a capital letter, designating the gene subfamily (>55% sequence identity), and the letter is followed by yet another number, to designate a particular gene. If there are allelic variants of the gene (>97% sequence identity), the number is followed by a lower case v (for "variant"), in turn followed by another number. Thus, when we cloned and sequenced a cytochrome P450 from the black swallowtail caterpillar *Papilio polyxenes*, it became known as *CYP6B1v1*—a member of family 6, to which belonged the first insect P450 cDNA to be cloned, but sufficiently different from said P450 to merit its own subfamily, 6B.

My biggest concern about presenting a paper at this particular meeting is that I wasn't sure how to pronounce "CYP6B1." Around the lab, we called this enzyme "sip-six-bee-one" but I really wasn't certain this was *de rigeur* among those in the know. I was exceedingly relieved when the first speaker of the day at West Lafayette, Paul Ortiz de Montellano, a highly respected figure in the field from University of California at San Francisco, began talking about "sip-1A1" and other mammalian P450s. My relief lasted only until the beginning of the second talk, when that speaker stood up and started describing his work on cytokine-mediated inhibition of "sipe-17" expression. For the record, the third speaker referred to his family of proteins with the descriptor "see-wye-pee." I confess that, by that point, I was too distressed to listen to the fourth talk at all.

BUZZWORDS

While acronyms have probably always been with us in biology, they've kind of been getting out of control lately. Time was when acronyms were more or less the exclusive province of electrical engineering. Remember SONAR? And LASER? But ever since the genetic material was discovered to be something with the exceptionally unwieldy name of deoxyribonucleic acid, biologists have gone in for acronyms (such as the classic "DNA") in a big way. Molecular biology is without doubt the most acronym-intensive area of contemporary biology. This field not only has acronyms, it has synonyms for acronyms. The external non-transcribed spacers (ENS) between ribosomal RNA genes, for example, are also known as the IGS, or intergenic spacers, *and* as the NTS, or non-transcribed spaces. For that matter, this field has acronyms made up of other acronyms. In *Drosophila*, some proteins contain a 270-amino acid motif that appears to facilitate self-association. When this motif was found in the period gene product (PER), in the aryl hydrocarbon nuclear translocator (ARNT) (a component of the dioxin receptor complex), and in the single-minded gene product (SIM), it was only natural to refer to it as the PAS domain (from PER-ARNT-SIM). The world hasn't seen this kind of acronym use since the days of the KGB in the USSR at the height of the Cold War.

I think excessive acronym use might be one of the reasons I am often slightly uncomfortable when reading articles dealing with molecular biology. There are always so many CAPITALIZED acronyms in the text—reading these WORDS in big letters makes it SEEM like the author is very ANGRY for some reason. Take, for example, the issue of *Science* from September 22, 1995, which happens to be sitting unfiled on my desk at the moment (along with other scattered issues of *Science* going back to 1981, when I first moved into this office). The page called "This Week in Science" describes highlights of the issue. The highlights include reports on control of contraction in muscle cells by Ca2+ release by the SR (sarcoplasmic reticulum), on control of protein phosphorylation and T cell proliferation by the chemokine RANTES, on regulation of IFN (interferon)-induced transcription by MAP (mitogen-activated protein) kinase (making it a STAT, or signal transducer and activator of transcription), on PSD-95 (a postsynaptic density protein), which interacts with a type of NMDA (N-methyl D-aspartate) receptor, and on the role of nAChRs (N-acetylcholine receptors) in the CNS (what nervous system?). The acronyms on the page seem to be shouting out for attention, STAT! These are not mere pleas for attention, they are RANTES!

The trend is clear—acronyms will not only not go away, they will infiltrate all areas of biology, including entomology. They've long been entrenched in insect physiology (e.g., the source of cellular energy, adenosine triphosphate, ATP, formerly known as TPN). Insect systematists use them with increasing frequency, not only to describe the methods they use to obtain data (e.g., SDS-PAGE, RFLP, RAPD), but also the methods they use to analyze their data (PHYLIP, PAUP). Systematists in fact may have gotten the whole acronym ball rolling centuries ago by adopting the convention of abbreviating the genus name when referring to a species. Nowadays, the bacterium *E. coli* and the nematode *C. elegans* barely have first names anymore, although for whatever reason *Drosophila melanogaster* has resolutely defied abbreviations. Even ethology has heeded the call, what with the Evolutionarily Stable Strategy generally referred to as an ESS.

About the only discipline in which acronyms are truly unusual is ecology. There, they are certainly few and far between. There are a few in population biology that I confess have long baffled me—why, for example, is "carrying capacity" designated "K" instead of "c" or "CC"? And the intrinsic rate of growth is known rather too concisely as "r" (pronounced "little are")—why, historically, was it never IRR? Why lower case r in the first place? Was it an uncommon fear of attracting attention?

Acronyms in ecology also don't seem to have much longevity (sometimes designated lx in life table analyses), either. My UIUC colleague Gilbert Waldbauer was successful back in 1968 in introducing a panoply of acronyms for discussions of ecophysiology. In a widely cited review of gravimetric estimates of nutritional performance, Waldbauer established the parameters ECI (efficiency of conversion of ingested food), ECD (efficiency of conversion of digested food), RGR (relative growth rate), RCR (relative consumption rate), and AD (approximate digestibility). But the future of these acronyms is in doubt, thanks to several recent publications by statistically minded biologists questioning the calculations upon which the acronyms are based. If these critics prevail, no longer will insect ecology students have acronyms to worry about while studying for midterm exams; knowing what ECD stands for will no longer mean the difference between an A and a B.

This undue reserve acronym-wise may put ecologists at a significant disadvantage in terms of competing in today's scientific world. I would argue it's a bad time for anyone to start expunging acronyms from their discipline. I can't help noticing that the allocation of federal research dollars to biological subdisciplines appears to be highly correlated with acronym use. Molecular biology, e.g., receives the lion's share and ecology, systematics, ethology, and other acronym-poor programs are suffering deep cuts. Whether this relationship is causative I can't really say. I do feel compelled to point out, though, that it wouldn't be inconsistent behavior from agencies widely known by their initials, NSF, NIH, USDA, and DOE. Coincidence? I don't think so. In fact, as an insect ecologist, I think I will devote some time developing some acronyms as soon as possible (ASAP), before the next request for proposals (RFP) is issued by the National Science Foundation (NSF).

"Quite without redeeming quality?"

Y ou might think that, after 20 years as an independent professional scientist, I'd be pretty much inured to criticism. After all, the review process is intrinsic to the conduct of contemporary science. Grant proposals are scrutinized by panels and ad hoc reviewers, manuscripts are criticized by reviewers and editors, and even course lectures are evaluated at the end of the semester by undergraduate students. I'm certainly no stranger to negative remarks, either—I've had my share over the past two decades. I actually keep all of my reviews, favorable and unfavorable, in a file cabinet in my office (they now fill two complete drawers) and read through them again every now and then. Even favorable reviews tend to have a few critical remarks, and it's almost always the case that even the most negative reviews I've received have had some merit to them, some constructive comments, no matter how inelegantly these comments have been phrased. The one that stands out as being particularly brutal was a review I received for a manuscript submitted to *Apidologie* (and eventually published, after extensive revision, in another journal). In the course of a single paragraph, this reviewer managed to use the words "inadequate," "naive," "superficial," and "farcical," as well as the phrase "quite without redeeming quality." Mercifully, most people tend to be a little more circumspect in their choice of adjectives. Conservatively, I can say I've received more than a hundred bad reviews over the course of my scientific career to date, counting just manuscripts and grant proposals. I console myself with the knowledge that not even Michael Jordan made every basket he shot for, either, and I regard every bad review as a learning experience.

Notwithstanding all of my experience in this context, I confess I was completely unprepared for a bad review that I received in 1998. Ironically enough, the review was directly related to my "Buzzwords" essays. In a fit of egotism, I thought that it might be a good idea to collect all of the columns I had written for *American Entomologist* and submit them to a publisher, in the hope of converting them into a book. Such an idea is hardly radical—a number of columnists, even those writing on scientific subjects, routinely publish their essays in book form (Steven J. Gould comes immediately to mind). Because I knew that Cornell University Press (CUP) has a history of publishing such works, and because I ran into Peter Prescott, the CUP acquisitions editor, at the Entomological Society of America meeting in Nashville in December 1997, it seemed like a good idea to bundle together the columns and send them off to Cornell University Press for appraisal. As all good editors should, Peter Prescott sent the package out for review. The first review came back quickly and was quite positive; I figured this was going to be easy. The second review, however, completely threw me. The reviewer, self-described as an entomologist who had taught and conducted research at a land-grant university for 40 years (and therefore eminently entitled to an expert opinion), was breathtakingly unimpressed with "Buzzwords." He described some of the essays as "inane" and apparently found references to "orifices, sex, and other basic bodily parts and functions" unappealing. "Little conceptual matter of any depth" was another phrase that appeared in the review, threatening to displace "farcical" as the most depressing criticism I've ever received.

I couldn't even take solace that this opinion might be an exceptional one. This reviewer, in a frenzy of thoroughness, polled a group of his or her entomological colleagues (also eminently entitled to expert opinions) and found, of 19 who knew of the columns, three who did not like them at all. Of those who admitted to liking the columns, none were reported to be "ebullient" in their praise. In contrast, those who did not like the columns were free with their use of adjectives and offered up words such as "trite," "weird," and "juvenile" to describe them.

Maybe ego is to blame, here, but it really never occurred to me that people might detest these essays. I always figured that a few people would ignore them, but I never thought they might be utterly despised. Reading this review was like discovering, in the middle of a prestigious lecture in a packed hall, that my pants zipper was undone and that audience members who were laughing were not marveling at my fine sense of humor but rather at my choice of undergarments that morning.

In short, after reading this review (over and over again) I became acutely self-conscious and acquired a major case of writer's block. The reason for the writer's block was that, as always, this negative review contained a considerable amount of truth. When I write, I do tend to go into orifices, as it were, with greater regularity than the vast bulk of entomological writers (although, if anything had "depth," one would think orifices would). My sense of humor

Getting the bugs out of the text

does on occasion slip from lofty to low-brow. I soon began to obsess over every word and wonder how I might recognize "trite" or "inane" whenever it crept into my writing. But after a couple of months, it struck me that the criticism was a little unfair. The humor may well be juvenile, but I happen to know it's not nearly as juvenile as it could be. As a matter of fact, there are several subjects I've refrained from writing about expressly because they seemed to *me* to be too juvenile. And, just to prove the point, I'll share them with you now.

BUZZWORDS

Here are four ideas I personally rejected as being too trite, juvenile, or tasteless to follow up on, even before explicit recognition of my shortcomings was thrust upon me.

1. According to Partridge (1974), there is an assortment of colorful turns of phrase that metaphorically associate insects with the act of masturbation, including "to box the Jesuit and get cockroaches" and "to gallop one's maggot."

2. In an article titled "Perfectly inflated genitalia every time," Nolch (1997) describes the development of the vesica everter, or "phalloblaster," developed by I. W. D. Engineering and the Australian CSIRO Division of Entomology. According to Dr. Marcus Matthews, quoted in the article, "The vesica everter inflates the genitalia with a stream of pressurised absolute alcohol which dehydrates and hardens the genitalia. . . . They then remain inflated like a balloon which never goes down." As a point of clarification, I should mention that the phalloblaster was designed for use only with insect genitalia.

3. Dangerfield and Mosugelo (1997) in an article titled "Termite foraging on toilet roll baits in semi-arid savanna, South-East Botswana (Isoptera: Termitidae)," describe a method for surveying termite abundance with the use of "single-ply, un-scented toilet rolls" as bait. According to the methods, each role is "placed in an augured hole," not to be "even" with the soil, or to be "level" with the soil, but to be "flush with the soil."

4. Following up on earlier reports that the yellow fever mosquito *Anopheles gambiae* is highly attracted to the odor of Limburger cheese, which bears a strong resemblance to the odor of (some) human feet, Knols et al. (1997) were able to produce a synthetic Limburger cheese (or foot) odor by characterizing the

chemical components of Limburger cheese headspace. Again, as clarification, cheese headspace is the volatile milieu immediately surrounding the Limburger, not the empty seat next to a guy wearing a funny hat at a Green Bay Packers game.

Any one of these topics would have made for an incredibly tasteless column, and, up to this point, I have restrained myself from succumbing to the temptation to get really sophomoric. Truth be told, though, it was more than just good taste that led to this restraint. I really don't think I could stand reading the reviews I might get if I actually went ahead with one of these ideas.

Flintstones 101

As the result of an accident of birth, I belong to a generally overlooked demographic group; born in 1953, I am among that elite group of scientists who were the first to grow up watching television. In 1948, television sales had surpassed sales of radios, and by 1955, when I was two years old, close to 2/3 of all American households had a television set. And I did watch a lot of television as a child. Among the more memorable of the programs I watched was "The Flintstones," which originally aired from 1960 to 1966 and is now still seen on cable stations almost around the clock in reruns. The Flintstones was something of a record–setter at the time; it was the first animated sitcom, it was the first animated 30-minute show, and it was the first animated series to feature human characters. I'd like to think I recognized it for what it was, which was an animated version of the sitcom "The Honeymooners." Evidently, though, many of my contemporaries apparently believed that they were watching a documentary. In a survey of science literacy conducted by the American Museum of Natural History in 1994, some 35% of adults polled believed that prehistoric humans coexisted with dinosaurs and another 14% thought it might be a possibility. Experts other than William

Hanna and Joe Barbera believe that dinosaurs went extinct at the end of the Cretaceous, some 65 million years ago, whereas protohominids don't even appear in the fossil record until, conservatively speaking, 10 to 20 million years ago and true hominids not until about 4 or 5 million years ago.

Sixy million years is more than a minor detail but it's just one example of the problem of science literacy in the U.S. It's entirely likely that the 49% of adults surveyed by the American Museum of Natural History had at some point in their lives heard that dinosaurs and humans didn't cohabit. Most likely they heard it in school—I remember distinctly my fourth grade teacher, Mr. Parker, discussing the issue—but for whatever reason it didn't register. It's a safe bet, however, that the majority of those very same adults could recognize the theme song from "The Flintstones" if not sing along to it in its entirety.

So why is it that science is so hard to remember? One possible reason is that it's boring—or at least it's perceived by the vast majority as such. There's no question that most find it boring in school. Innumerable studies have revealed that students find the way science is taught in school to be insufferably dull. Recent examples include:

1. An Office of Technology Assessment study conducted in 1988, called *Educating Scientists and Engineers; Grade School to Grad School*, in which high school students polled by Harvard's Educational Technology Center agreed overwhelmingly that science is "important but boring."

2. A recent study of *Why Students Leave* undergraduate science majors, in which authors Seymour and Hewitt report that the proportion of undergraduates remaining in science majors is

substantially lower than the proportion remaining in social science or humanities majors, and the principal reason these students switched majors (mostly out of science altogether) was that they were "turned off" or bored by science.

It's unquestionable that there are problems in the classroom and there are major efforts under way to reform science education in schools. There have been for years, in fact; headlines from 10 and 20 years ago recognized the same crisis that is freely discussed in the press today. The repeatedly poor performance of American eighth graders in the Third International Mathematics and Science Study (TIMSS) relative to other eighth graders in 40 other countries invariably stimulates rounds of discussion in science and education journals alike (only 13% of U.S. students scored in the top 10% of 300,000 students in 1996). But while it's necessary to reform science education in schools, it's not sufficient. The reason it's not sufficient is that the amount of exposure to science that average Americans receive in school is literally an eyeblink over the course of their lives: a few hours a week in primary school and, over their four-year college careers, maybe, if they're lucky (or unlucky, depending on perspective), one course of 44 lectures in biological science and one course of 44 lectures in physical science. That's six hours out of a minimum 120 (5%). For biology, that translates to 35 literal hours of lecture (no labs required) over an entire undergraduate career, not counting classes missed due to rainy days, oversleeping, days that are too nice to go to class, the Friday before spring break, the Wednesday before the Friday before spring break, and other exigencies of undergraduate life.

Where, then, do these students get exposed to science, out of school and after graduation? The chief source of science information over the lifetime of the average American is, most likely, television. Unlike mandatory school attendance, which has held steadily constant for many years, the number of hours spent watching television has been steadily increasing. The three or four 50-minute periods per week of science education required by enlightened school districts at the elementary and middle school level pale into insignificance compared to the average 12-14 hours of television-watching per week these same students engage in. Once the semester ends, and once students graduate, it's all television, newspaper, and other media, except for perhaps the very occasional public lecture.

It's not that learning from television is a bad thing—I've learned a lot from television. I learned, e.g., how to spell "polo pony" from a memorable Flintstones episode where Fred plays scrabble. But there's a problem. Science on television is boring. Joe Queenan, a TV commentator for *TV Guide*, the most widely read periodical in the U.S., once proposed a new rating system for TV (April 25, 1997), with "TVB" as a viewer advisory for boring shows—such as science documentaries and speeches by Vice President Al Gore. There's definitely a rut out there. I surveyed titles in the Teachers Video catalogue, a compilation of among the best science documentaries available on video, and found that almost 90% of titles are either nouns + prepositional phrases ("Kingdom of the Sea Horse"), single nouns ("Tornado!"), or adjective + noun ("Invisible World"). It's likely that no other form of entertainment displays such depressing uniformity. Titles from other forms of film, e.g., are rife with alliteration, puns, literary references, and puncutation marks other than colons. Take soft-core porn film titles; a quick review of a video catalogue reveals alliteration ("Beach Babes From Beyond"), allusion ("Oddly Coupled,"

"Midnight Ploughboy," and "Thigh Noon"), and puns ("Dracula Sucks").

Why should anyone bother about reaching out to the masses? First of all, there's a vast potential audience out there for a good production, and there is a lot of money to be made. Carl Sagan's "Cosmos" series attracted more than 500 million viewers over its run. With those audiences, sales potential isn't too shabby—the Stauffer brothers, e.g., sold close to $20 million worth of nature videos in a three-year period and the David Attenborough "Life on Earth" series generated more than $20 million in revenue in its day. That's not even considering the potential for merchandising. "Magic School Bus," a PBS program aimed at teaching science to kids, managed to partner with McDonald's to put science into Happy Meals. Of course, the lines for Ms. Frizzle and her entourage weren't nearly as long as they were for Teeny Beanie Babies, but it's a start.

There's a more important reason, too, for becoming involved in the production of high quality accurate AND entertaining science programming. I think everyone recognizes that in a democracy everyone is better off if the majority of voters are well informed. An increasing number of laws and referenda deal with issues that involve science. Poll after poll demonstrates that Americans feel it's important to be well-educated in science. The American Museum of Natural History study (1994) revealed that 76% of the 1,200+ individuals polled admitted to enjoying learning about science for its own sake. And a 1991 SIPI/Harris Poll showed that the level of interest in the general public runs as high as 80% for certain issues of science or policy. And a 1996 study conducted through the Chicago Academy of Sciences on Public Understanding of Science and Technology demonstrated the broad gap between science interest and science knowledge here in the U.S., compared with most other nations.

Most people recognize that, to deal with complex issues, they need information. For the most part, they're willing and eager to be educated and it's important for scientists to give them what they want and need, or to work with media professionals to do it. Otherwise, we all run the risk of allowing tabloid newspapers to determine science policy and living with the consequences. After all, do we really want to count on"Weekly World News Readers" to "Shrink Holes in Ozone Layer—With Their Minds!"?

A word to freshmen

In late summer a massive human migration begins, one without parallel in the natural world. From all over the country, adolescents and young adults depart their family homes and set off for college campuses of all descriptions. They arrive on these campuses burdened not only with the necessities of college life—pillows, desk lamps, and hot plates—and the accoutrements of higher learning—textbooks, personal computers, pencils, and erasers—but also with an amazing number of preconceived notions about what college life will be like. A distressing number of these preconceived notions focus on a group of people that is, to incoming freshmen, something of an enigma—college professors.

Incoming freshmen can hardly be blamed for having a few odd ideas about what exactly it is that a professor does. The vast majority, in all likelihood, have never met one. It's not like there are professors living in most cities or towns in the U.S. If you look in the Yellow Pages of whatever town you're from, you almost certainly won't find a listing for "Professors," like you do for auto mechanics, taxidermists, dance instructors, veterinarians, taxi drivers, and other overtly useful people. In fact, even here in

How an entomologist sees science

Champaign-Urbana, a town crawling with professors, you won't find "Professors" in the Yellow Page between "Process Servers" and "Prosthetic Device," where you might expect to find them (alphabetically, that is).

So most freshmen arrive on campus without any personal experience to guide their impressions. It's not that they lack impressions. Unfortunately, the one source of information people are most likely to use in formulating their image of a professor is television and I'm not certain that television is the best source of information on which to build any sort of impression. A quick check around the dial illustrates the point. If, for example, your tastes run to comedies, you may have the impression that professors are like Professor Julius F. Kelp, as portrayed by Jerry Lewis in the 1963 film, "The Nutty Professor"—a socially inept, inoffensive fellow who uses his powerful intellect and scientific expertise to perfect a chemical formula that can turn himself into a lounge singer—the suave and sophisticated Buddy Love, bon vivant and man about town who is irresistible to women, including female students in his classes (the fact that he ends up dating them would likely have him hauled up before his Dean on harassment charges today). If not Buddy Love, you might have the impression that college professors are just like Professor Ned Brainard (Fred MacMurray), the absent-minded professor from the 1961 film of the same name—another socially inept inoffensive fellow who develops a chemical formula for flying rubber that he uses to help his college's basketball team win The Big Game. Now, I'm not exactly an authority on the subject but my guess is that the NCAA would probably frown on this sort of thing if they found out about it.

Sci-fi fans can base their impressions on the 1958 cult classic,

"Monster on the Campus," about Dr. Donald Blake (Arthur Franz), well-intentioned scientist and researcher at Dunsfield University who, by virtue of exceptionally bad laboratory technique, inadvertently gets exposed to the blood of a Carboniferous era fish (through some amazing plot contrivances, he ends up smoking it) and turns into a primitive raging violent humanoid mutant. This film, by the way, was once reviewed by the *New York Times* as "summa cum lousy." If you're a fan of adventure films, then you can see Harrison Ford as Indiana Jones, the anthropology professor who, over the course of his academic and movie career, found the Ark of the Covenant, saved the Holy Grail from Nazis, and rescued people from the Temple of Doom sustained only on occasional meals of monkey brains and scarab beetles. Or maybe you can see Sam Neill as the paleontologist in "Jurassic Park," who saves the world from rampaging velociraptors, presumably in between cataloguing fossils and grading term papers.

So, based on your television viewing habits, you might think professors are bumbling misfits, superheroes, or twisted deviates. And, whatever other category they belong to, at least in the movies, they are also all white males. The reality is, to be perfectly frank, a lot duller in almost every respect. A demographic breakdown of the UIUC faculty quickly dispenses with the notion at least that professors are invariably aging white males. While the ratio isn't exactly 50:50, as it is in the public at large, it's a lot better than it was back in Julius Kelp's day. And while most of the faculty aren't exactly Generation X-ers, neither are they old enough to have accompanied Teddy Roosevelt and his Rough Riders on the charge up San Juan Hill. Nor is the UIUC faculty all-white, either; there are African-Americans, Latinos, Asians, and Native Americans among the ranks.

How an entomologist sees science

Demographic data obtained from censuses don't really provide much information on the number of superheroes or fiends on the campus, nor do they shed much light on degree of social ineptitude. Data that could be informative in this context are very hard to collect and I was unsuccessful in obtaining these data from the campus; after all, it's not as if the central administration can pass out questionnaires with questions like, "I am a dweeb—yes, no, uncertain." In an attempt to fill the gap at least a little, I collected some information from the faculty in my own department, the Department of Entomology in the College of Liberal Arts and Sciences.

—Each faculty member on average has a single spouse and 1.88 children; cats are the most popular pets (and, just to update this survey, which was conducted some time ago, I regret to say that the untimely death of Flounder, the guppy, has reduced the number of faculty with fish from three to two).

—Hobbies run the gamut from stamp and coin collecting (hobbies Julius Kelp would undoubtedly have appreciated) to decidedly less dweeb-like tae kwan do and wind surfing.

—In terms of musical tastes, while admittedly there is a conspicuous absence of contemporary groups among the favorites (no Barenaked Ladies or Nine-Inch-Nails yet), variance is very high and runs from sublime (Chicago Symphony) to ridiculous (Herman's Hermits).

If these data do show a pattern, it's that the faculty, even in a tiny department, are a variable lot. With such variability, generalizing is a risky business. It's too often been the case that incoming freshmen have assembled bits and pieces of information gathered from media and from friends and brought with them prejudices that lead them to major misconceptions about the college professor, here as well as at other institutions. I'd like to take this opportunity to reveal the most common of these misconceptions for what they are.

—POPULAR MISCONCEPTION NO. 1. *Professors prefer to think of their students as numbers rather than as individuals with personalities and even names. Professors probably don't even name their own kids, they just refer to them by the last four digits of their social security numbers.*

It's true that there are big classes at public universities and, for logistical reasons, numbers are necessary. And it's true that most professors don't make an effort to learn the names of all students in classes with 500 or more of them. But I don't know anyone who doesn't want to get to know his or her students; oftentimes that's difficult to do because students are so reluctant to approach their professor, even to ask a question. In a small class, odds are very good that your professor will learn your name, and maybe even remember it years hence. In a large class, it's hard to say—but, as a general rule of thumb, whether or not a professor learns to recognize you, remember you, and address you by name depends a great deal on how much effort you put into the class. I would like to add, on behalf of my colleagues, that it's very hard for us to learn to recognize you and address you by name if you never come to class.

—POPULAR MISCONCEPTION NO. 2. *Professors at a prestigious universities like to conduct Important Research; this Important*

How an entomologist sees science

Research has no useful purpose other than to ensure that the Professor conducting it gets tenure, at which point he or she no longer needs to do it anymore.

Again, activities of the faculty here dispel this notion quickly; research conducted by UIUC faculty, as a case in point, has affected the lives of an enormous number of people. The light-emitting diodes in digital clocks, transistors in transistor radios, Illini supersweet corn at July 4th picnics, the soundtracks on movies at the movie theater, the insecticide in the flea powder you sprinkle on your dog, the software you use to surf the Internet, the biodegradable packing material in the present your grandmother mailed you on your last birthday—all were based on research conducted by UIUC faculty. Among the most satisfying aspects of research in any field is to see your work cited and used by others, not buried and forgotten in obscure journals that not even the University of Illinois Library can afford to subscribe to anymore. UIUC is not by any means unique—just about any public university can boast of an equivalent list of accomplishments. I remember a high point in graduate school at Cornell being the moment I realized that I was standing in a lunch line behind the man who invented the turkey hot dog.

—POPULAR MISCONCEPTION NO. 3. *Professors lead such empty and meaningless lives that their sole source of pleasure is taking points away from students on exams. What they do with all of these points they take away is a baffling mystery, although it's been suggested that they store them in huge vaults and go count them at the end of the semester.*

In point of fact, most professors don't like taking points off a student's exam. If you think about it, it requires far more effort to take points off than not to take points off. The easiest kind of exam to grade is one that's perfect. Not only is it easier, it's also a lot less depressing. I personally rejoice when a student achieves a perfect exam score in a course I'm teaching because I can take that score to mean I've been effective in transferring information to that student. So students do everyone a favor when they study and get as many answers right as they can.

—POPULAR MISCONCEPTION NO. 4. *Professors are intentionally boring. They would have to be working at it to be as boring as they actually are. Professors who are interesting get fired from the university because they have violated the first rule of academic classroom performance—if it's not boring, it can't be educational (like medicine, which can't be effective if it doesn't taste bad).*

I have to admit that you will encounter some professors whom you will find boring. But then again, odds are good that other students in the same class may find that same professor absolutely fascinating. There are as many styles of teaching as there are faculty members and, just like you may not like all 31 flavors of Baskin-Robbins ice cream, you might not like everybody's teaching style. Most of the faculty I know try very hard NOT to be boring, in the classroom or in their lives outside the classroom. Imagine if you were having a conversation with someone and before you could finish what you were saying that person started snoring—I can assure you, there is nothing more demoralizing than looking out and seeing even one student asleep during your lecture.

How an entomologist sees science

There are many more misconceptions, but to list them ad nauseum would just belabor the point—the point being that, through your academic career, you should try to keep an open mind about professors. We're as different from each other as each of you is from anyone else in the world (although an exception may have to be made for sibling professors who are identical twins). There is one thing, though, that we all have in common— at one point in our lives, we were all college freshmen. It may have been a long, long time ago for some of us, but we've been there— there are some things one never forgets. So get to know us if you can—you might be amazed at what you find out. Some of my best memories of my college days, believe it or not, involve the professors I met there—the ones who taught inspiring courses, who took a personal interest in my career, and who in countless ways helped me become, for better or worse, the person I am. I wish you many such memories to come. By the way, don't worry—you won't be tested on this material four years from now . . . and yes, the final is cumulative.

Holding the bag

In his masterful autobiography, *Naturalist*, the great evolutionary biologist Edward O. Wilson speculated that physical limitations can determine the course of a life. In his case, a painfully close encounter with a pinfish during a childhood fishing expedition left him with a left eye that couldn't focus at long distances; not coincidentally, he devoted his career to the study of ants and other small creatures that require magnification for close observation. I am an enthusiastic subscriber to this theory, because I know of at least one physical infirmity that I possess that has influenced the course of my own career. This particular limitation did not arise as a result of a dramatic conflict with nature; rather, it seems to have been an accident of genetics. I've spent my entire research career to date working on organisms that live within walking distance of the laboratory because, since before I can remember, I've been exceedingly prone to motion sickness.

Motion sickness, as you most likely know, results from a genetic inner ear imbalance. Everyone experiences motion sickness at some point in their life—deep sea fishing or shuttle launches can usually unsettle even the strongest of stomachs. My threshold is a lot lower. Not only did I get car-sick with depressing regularity as a child, I also, on various occasions, got elevator-sick, train-sick, bus-sick, ferry-sick, rowboat-sick, and, on one memorable occasion, bicycle-sick. There are times when I swear that continental drift is making me motion-sick. A motion-sickness bag was always at the ready in the glove compartment of my parents' cars throughout my entire childhood; to this day, I make sure I carry one with me whenever I know I'll be in a moving vehicle for more than ten minutes. I may forget to pack my lecture notes or bring the wrong set of slides when I leave for a trip, but I always have my bag with me (not that it's a big help at seminar time when I have brought the wrong slides).

Despite parental reassurances that I would outgrow this unsettling tendency, it hasn't happened yet. As a consequence, many aspects of my professional life have been affected. The traveling I do to conferences and seminars is a constant source of apprehension and, to some degree, aversion. Most recently, I gave a twenty-dollar tip, over and above a 20% tip, to the limo driver who took me from Oxford, Ohio, where I'd given a lecture at Miami University the day before, to the Cincinnati airport because I threw up in his beautiful, brand-new, boat-like Lincoln Town car. Colleagues think I'm an exercise fanatic because I prefer to walk a mile-and-a-half from the hotel to a conference center; they don't know that subway trains and taxis in stop-and-go city traffic are more than my system can handle. And, when the Entomological Society of America had its annual meeting in the Las Vegas Hilton Hotel, I declined to join my colleagues in the Star Trek Experience not because there was an interesting symposium I didn't want to miss but because I thought simulated warp drive would push me over the edge. As for choice of field sites, no tropical rainforests for me—my field sites are for the most part less than fifteen minutes away from the lab. And even in my office, susceptibility to motion sickness affects many daily decisions; I don't have a screensaver on my computer because the sight of flying toasters turns my stomach.

As a biologist, I can't help wondering what nasty trick of evolutionary physiology has forged a connection between perceived aberrant motion and vomiting. As far as I know, there has only been a single paper published on the subject. In 1973, Michel Treisman published a paper in *Science* titled, "Motion sickness: an evoutionary hypothesis." Treisman notes that motion sickness is triggered by conflict among the sensory inputs by which humans determine their location in physical space: the visual frame of reference provided by input from the eyes, orientation of the head as indicated by vestibular receptors in the ears, and relationships among body parts as indicated by non-vestibular proprioceptors. Inconsistencies among these inputs can be realigned to some degree, by head or eye movements, or compensatory movements of trunk and limbs. Modern life, however, provides inconsistent inputs that exceed the adaptational capacity of the system. Traveling at speeds in excess of 60 miles per hour (or, even more dramatically, 600 mph in an airplane) can cause sensory conflict that is not easily resolved by realignment of the head or alteration in patterns of eye movement.

As to why such conflict should lead to vomiting, Treisman suggests that, in our evolutionary history, the stimulus most likely to produce discordance among the major sensory inputs is ingestion of a nerve poison, or neurotoxin. Plants are rich sources of substances that disrupt normal neurological function (as I write this at 3 a.m. in an agitated state directly attributable to the consumption of caffeine-laden Dr Pepper at lunchtime). An emetic, or vomiting, response to a mismatch between inputs is adaptive in this context in that it provides the body with a mechanism for removing neurotoxins from the system. The author goes on to suggest that the malaise and "marked depression" that often accompany motion sickness may be adaptive as well, that in our evolutionary past such aversive responses would serve to reduce the probability of future encounters with the source of the nausea-inducing material. I expect there is some truth to this notion; if I ever go back to Miami University I definitely will check to see if there's another limo service in town.

Treisman also reports that humans are not alone in their tendency to experience motion sickness—monkeys, horses, sheep, some birds, "and even codfish" (but "not rabbits or guinea pigs") are prone to motion sickness. This raises the question as to how one recognizes motion sickness in nonhuman species. In humans, the principal manifestations are pallor, sweating, and vomiting. I don't know if fish can sweat, but I'm certain that horses can't vomit—that's one reason they are so prone to colic, or intestinal difficulties. And I'm not sure how one determines whether a sheep looks pallid or not. Assuming motion sickness can be recognized when it appears, I'm not convinced that the taxonomic distribution of motion sickness is supportive of the hypothesis. Among the nonhuman model systems used for the experimental study of motion sickness are carnivores such as cats, dogs, and ferrets, omnivores such as rats, and an insectivore known as *Suncus murinus*, all species unlikely to ingest plant toxins (although some insects produce neurotoxic defense secretions and carrion may contain some nasty toxins as well).

I find this literature fascinating, and I would love to explore the subject in greater depth, but here's where physical limitations enter in again. I find I can't read statements such as "ingestion of warm yoghurt intensified nausea during Coriolus stimulation in a rotating chair" (Feinle et al., 1995) without slamming the paper down and running to the bathroom. I happily leave the study of motion sickness, and the warm yoghurt, for that matter, to those with more adept vestibular sensory systems and stronger stomachs than I possess.

Kids Pour Coffee on Fat Girl Scouts

We're all products of the era in which we grew up. That I came of age during the 1960s and attended college in the 1970s I suppose is reflected in my dress and appearance—with jeans (which my high school didn't allow girls to wear until senior year) and long hair (which my parents wouldn't allow me to have until I moved away to college). That I still dress like a college student isn't so much a smoldering rebellion against a long-defunct establishment as much as it is inertia. My background and upbringing are also reflected in my teaching—specifically, in the mnemonics I use and share with students.

Mnemonic devices are memory boosts—those rhymes, poems, slogans, or sketches that help people remember certain facts. The word derives from the Greek "Mneme" for memory and evokes Mnemosyne, the goddess of memory and mother of the muses. Supposedly, the literary use of mnemonics goes back 2,500 years, to Simonides the Younger, who promoted their use in the fifth century BC. Probably everyone has been exposed to mnemonics at some point in school, if only, perhaps, in elementary school spelling classes in order to learn that "I goes before E except after C or when sounding like (A) as in "neighbor" and (weigh)." Generations of schoolchildren are thus doomed to throw their hands up in despair upon reading that "neither foreigners nor financiers forfeited their weird heights leisurely." They may also be driven to give up protein from heifers for good. But, because memory boosters must themselves be more memorable than the facts they're boosting, they are often most effective when they're not just catchy rhymes but rather are customized to reflect the time and place of the rememberer.

One example from high school mathematics serves to illustrate this point. Trigonometry is the mathematical study of relationships among arcs and angles defined with respect to a circle; trigonometric functions allow the calculation of angles from known sides of angles and vice versa. If the circle is divided into quadrants, the trigonometric functions associated with triangles described with respect to the circle will be variously positive or negative, depending on the quadrant. In the first quadrant (equivalent to all angles between 12 and 3 on a clock face), all of the functions (sine, cosine, secant, and tangent) are positive. In the second quadrant (between 12 and 9), only the sine is positive; in the third quadrant (between 9 and 6), it's the tangent that's positive, and in the fourth quadrant (between 6 and 3), the cosine is positive.

In Williamsville High School, near Buffalo, New York, in 1970 it was important to remember these functions (although I must admit I can't remember why, other than to get a good grade on a trigonometry exam). The mnemonic device that our teacher shared with us was "Albany State Teachers College"—All, Sine, Tangent, Cosine. This was a fairly effective mnemonic—I certainly remember it, 30 years later, although I've long had a sneaking suspicion that there never was a teachers college in Albany, New York.

But the point is that this mnemonic wouldn't have been particularly useful in West Virginia, where high school students couldn't be expected to have any connection to the state capital of New York. I don't know how schoolchildren in West Virginia remember the positive trigonometric functions. Some mnemonics texts suggest as an alternative "All Stations to Coventry," which I expect is British in origin and must assume some familiarity with train schedules in the United Kingdom—again, not particularly helpful for students in West Virginia or anywhere else in the United States.

Biology is absolutely chockablock with mnemonic devices, at least in part because there is so much to memorize. In plant taxonomy we remembered the families of trees with opposite, rather than alternate, leaves by thinking of a MADCAP Horse— maple, ash, dogwood, Caprifoliaceae (honeysuckle), and horse chestnut. Human anatomy contributes many old classics—in the case of the spinal nerves, literally classic. My mother taught me a mnemonic for remembering the 12 spinal nerves—"On Old Olympus' Towering Tops/A Finn and German Viewed Some Hops," for the olfactory, optic, oculomotor, trochlear, trigeminal, abducens, facial, acoustic, glossopharyngeal, vagus, spinal accessory, and hypoglossal nerves. Variants of course exist—one rude one asserts that the only individual viewing hops is a "Fat-Assed German" (Dr. Crypton, 1984).

Naughtiness seems to be rife among mnemonics—naughtiness is perhaps more memorable than, say, logic and good grammar. The more contemporary the mnemonic, the more likely it is to be if not obscene at least in bad taste. Many years ago, George Gamow developed a mnemonic for remembering the ten classes of stars, from hottest to coldest—"O, B, A, F, G, K, M, R, N, S" thus became "Oh, Be A Fine Girl, Kiss Me Right Now, Sweetheart." As mnemonics go, it's cute but hardly X-rated. By the time I got to college, things in the mnemonics world had heated up considerably. The teaching assistants I had in geology class were more than willing to provide us with a mnemonic for remembering the Moh scale of hardness—the progressive increase in hardness of minerals. The scale of ten included, in order, talc, gypsum, calcite, fluorite, apatite, orthoclase, quartz, topaz, corundum, and diamond. The mnemonic you typically see in books is "Texas girls can flirt and other queer types can do," which grammatically doesn't make a lot of sense. This lack of grammatical correctness hasn't diminished its popularity. It has been embellished, however; according to the version the teaching assistants taught us, the Texas girls were considerably friendlier and had moved well beyond flirting. For that matter, the embellishment seems entirely appropriate in a scale of hardness.

Perhaps it's not a general pattern—this accumulation I have of naughty mnemonics might simply reflect the fact that I went to school in the freshly liberated sixties and seventies. The mnemonic I share with undergraduates today for remembering the taxonomic hierarchy—Kingdom, Phylum, Class, Order, Family, Genus, Species—is a case in point. There are many variants—among them, "Kings Play Chess on Fridays, Generally Speaking," or "Kids Pour Coffee on Fat Girl Scouts," or "King Philip Came Over From Glorious Scotland." But the version I learned in mammalogy class in 1973, and the one I should probably stop teaching, is "King Philip Came Over From Germany, Stoned." I don't know which King Philip it was and whether he was more likely to come over from Scotland clear-eyed and sober or from Germany with the munchies, but my students don't seem to care. In fact, they usually get every question about taxonomic hierarchy correct on their exams.

I continued to acquire mnemonics throughout my undergraduate career, spanning the range from naughty to nice. There is a definite drawback to naughty mnemonics, though, in that they can be too memorable. I spent a summer in Fayetteville, Arkansas, in 1974 assisting in an ecological survey of Devil's Den State Park. As part of that survey, the team conducted a regular census of breeding birds, which we did by walking through a defined section of forest and recording and identifying bird calls—breeding birds defend their territories by singing and the number of calling males is a good indicator of the number of breeding pairs. We used the *Field Guide to Birds of Eastern North America* by Roger Tory Peterson to aid us in identifying the calls. For each bird species, the author provided a clever mnemonic phrase to aid in recognition—thus, the trill of a Carolina wren was rendered "Teakettle, teakettle, teakettle," and the unique call of the ovenbird as "teacher, teacher, TEACHER!"

All this was fine until some members of the team thought it would be a good idea to substitute obscene equivalents of the mnemonic phrases. These quickly supplanted the quaint "pee-iks" and "quarks" of Peterson. These phrases were certainly easily remembered and instantly recognizable and greatly assisted us in completing the breeding bird census quickly and efficiently. But to this day, every time I walk through the woods when the birds are singing, I still have the unshakable feeling they're actually screaming obscenities at me, and nature doesn't seem quite as peaceful and serene after you've been propositioned by a foul-mouthed scarlet tanager.

An o-pun and shut case

Ever since I acquired my first joke book (*Arrow Book of Jokes and Riddles*), I've had a peculiar affinity for puns. This particular comic bent was not shared by anyone in my immediate family, as I discovered every time I tried out a new one on whoever happened by. Nor was it shared by my orthodontist, Dr. Sidney Elfant. I was so nervous during my first visit to Dr. Elfant's office that I actually threw up in the chair, whereupon, in an effort to make me less apprehensive, Dr. Elfant hit upon the idea of asking me to come back next time with a joke to tell him. Being as compulsive then as I am now, I spent the next two weeks studying and analyzing jokes, trying to select an appropriate one and complete the task assigned. It seemed to have worked, since I didn't throw up during the second appointment, and it became a regular practice for me to come to each appointment with a joke. Unfortunately for Dr. Elfant, I had to wear braces for five years, so he had to endure quite a few jokes in the process. I couldn't help noticing that whenever I told a joke that relied on a pun for the punch line, I always ended up with considerably more dental wicks in my mouth—making it much more difficult to relate any additional jokes during the rest of the appointment.

This interest in puns continued through high school. In fact, my Advanced Placement English teacher, Mr. Stein, chose a verse from Samuel Johnson with which to sign my yearbook:

"If I were punishéd
For every pun I shed,
There'd be no puny shed
Above my punnish head."

In college, too, I was free to indulge in punning, but everything came to a grinding halt in graduate school. I learned quickly that puns and science are not universally regarded as logical partners. Professors ruthlessly and routinely rogued out any wordplay from term papers; editors rarely allowed one to get by. I think, in all of my 100+ published papers, I have succeeded in slipping only a single pun past an editor—it was a paper on furanocoumarin content of parsnip fruits and I thanked a colleague in the acknowledgments for "fruitful discussion."

Admittedly, it wasn't even a masterful pun that I managed to get in print. Such do exist; one I'm particularly impressed with is in a paper by L. A. Fuiman and R. A. Batty in the *Journal of Experimental Biology* on the effects of temperature-induced changes in the viscosity of water on swimming performance of Atlantic herring. Appropriately enough, the paper is titled, "What a drag it is getting cold: partitioning the physical and physiological effects of temperature on fish swimming." The part preceding the colon is a sly reference to the first line of a Rolling Stones' song called "Mother's Little Helper." I wonder whether it was Fuiman or Batty who thought of it, and I really wonder how they managed to convince an editor to let it stand. A close second in punderstatement would be the paper by T. M. Blackburn, K. J. Gaston, R. M. Quinn, H. Arnold, and R. D. Gregory published in 1997 on the relationship between abundance and geographic size range in British mammals and birds, appropriately titled, "Of mice and wrens." And there's the paper by L. Clayton, M. Keeling, and E. J. Milner-Gulland from 1997 presenting a spatial model of wild pig hunting in Sulawesi, Indonesia, titled "Bringing home the bacon." I wish I could have found such accommodating editors for my efforts at scientific punning.

My experiences with editors (indeed, with just about everyone) lead me to wonder what is behind a sense of humor. Puns are almost universally despised—proponents are far outnumbered by detractors. The pun has been called "the lowest form of wit" by humorists (as the bun has been called the lowest form of wheat by bakers). In the eighteenth century, James Boswell felt that "a good pun may be admitted among the smaller excellencies of lively conversation," but that sentiment isn't surprising considering he spent a lot of time with Samuel Johnson (see above). William Combe was less enthusiastic; in his view, a pun was "a paltry, humbug jest; those who have the least wit make them best." By the nineteenth century, Oliver Wendell Holmes was declaring that "people that make puns are like wanton boys that put coppers on the railroad tracks. They amuse themselves and other children, but their little trick may upset a freight rain of conversation for the sake of a battered witticism." And the opprobrium continues in the twentieth century. Fred Allen was reported to have said, "Hanging is too good for a man who makes puns; he should be drawn and quoted" and contemporary humorist Dave Barry wrote, "Puns are little 'plays on words' that a breed of person loves to spring on you and then look at you in a certain self-satisfied way to indicate that he thinks that you must think he is by far the cleverest person on Earth now that Benjamin Franklin is dead, when in fact what you are thinking is that if this person ever ends up in a lifeboat the other passengers will hurl him overboard by the end of the first day even if they have plenty of food and water" (Barry, 1988).

I can't help it—I still think they're funny. And a search of the biology literature fails to yield insights into why. It's not clear what the adaptive value of a sense of humor might be at all. Although early studies suggested that a sense of humor may mitigate stress, more carefully controlled follow-up studies failed to produce any evidence that a sense of humor moderates the impact of stress on physical illness (Porterfield, 1987). There are very few studies documenting a health-enhancing effect of humor. One study conducted by K. M. Dillon and colleagues in 1985 did document an increase in people's salivary immunoglobulin production after they viewed a humorous videotape. This may be suggestive of a beneficial effect; immunoglobulins help the body defend against infection. This finding also suggests a novel quantitative criterion for use in judging stand-up comedy competitions, although practical implementation (which would involve collecting spit from the audience and judges) may prove difficult.

This field of research presents daunting operational challenges in that there is no objective, quantitative way to measure someone's sense of humor. Psychologists use instruments such as the Situational Humor Response Questionnaire, which asks subjects to indicate how much amusement they might derive from common situations on a 5-point scale. This questionnaire thus is designed to measure "mirthfulness." There are also humor appreciation tests that measure a subject's ability to detect the incongruity between joke stem and punchline and to evaluate the appropriateness of a punch line and a joke, as well as tests that ask subjects to rate individually a series of items that are designed to be either humorous or neutral.

Depite the obstacles, considerable progress is being made at determining the biological roots of humor appreciation. Shammi and Stuss (1999) conducted a study of 21 patients with focal damage in specific areas of their brains, using a series of humor appreciation tests and joke and story completion tests. For example, the patients were asked to read the setup of a joke and then select among a choice of punchlines. The investigators found that damage to the right frontal lobe had the most profound effect on the ability to appreciate humor. Although appreciation of slapstick was undiminished, the ability to pick out the clever line was muted. Thus, in one joke, a man asks his neighbor if he'll be mowing his lawn on Saturday; the neighbor answers "yes," and what was generally regarded as the clever punch line was for the first man to respond, "Well, I guess you won't be needing your golf clubs." Patients who had sustained damage to the right frontal lobe preferred the outcome in which, after the neighbor responds, "yes," the first man steps on a rake.

The right frontal lobe is the region of the brain that integrates cognitive and affective information. Humor experts such as T. P. Millar argue that there are two elements of humor: bisociation, in which a joke leads from one associative plane to another, and affective loading, in which a tension is generated by the joke situation. Puns are almost entirely bisociation; affective tension is minimal. That's okay by me. The idea that two silkworms had a race that ended in a tie is funny to me. So is the notion that a termite can walk up to a bar and ask, "Is the bartender here?" Maybe people who think puns are funny have affective deficits. Or maybe there are just genetic differences in intensity of activity in the different brain regions that govern these functions.

I like the idea that there may be a genetic basis for why I think puns are so funny. I actually did a literature search to see if there might be a pun gene. I found several references in a literature database. There is a *pun* gene in *Drosophila melanogaster*, reported by Fahmy and Fahmy in 1958, but *pun* is just an abbreviation for puny; the mutation affects eye morphology and wing length, not any tendency to resort to wordplay. This is unfortunate in that there is at least one well-known pun involving *Drosophila melanogaster*, according to which "time flies like an arrow but fruit flies like an apple." The database turned up two other potential *pun* genes, but further investigation proved disappointing. Hammer et al. (1999), in *FEMS Microbiology Letters*, was about "the *pun*ABCD gene cluster from *Pseudomonas fluorescens*," a bacterium; Gomi et al. (1994) in *Neuroscience Letters* was about "the *Pun* gene encoding PrP in titter rats." Both of these references seemed suspect to me, although the one about titter rats clearly held a lot of promise. Fetching these references provided some unwelcome clarification. The paper by Hammer et al. (1999) was actually about the prnABCD gene complex (a pyrrolnitrin biosynthetic gene cluster) and Gomi et al. (1994) about *Prn* genes (prion protein genes) in zitter rats, not titter rats, zitter rats being a neurological mutant strain isolated in 1978 and characterized by tremulous motion, hind limb paresis, and pathological changes in the central nervous system. So both of these pun genes were actually typographical errors. I find that very amusing—but, then, I would, wouldn't I? I only wish I could have found ten such typos, none of which represented a true gene for punning—then I could report with all seriousness that, after a search for such a gene controlling word play, no *pun* in ten did.

Hand-me-down genes

There are times when being a biologist affects the way you look at life. And why shouldn't it? Biology is, after all, the science of life and when something goes wrong with a life the majority of biologists experience at least a passing curiosity about the underlying mechanism responsible for the problem. Lots of things went wrong when my daughter, Hannah, was born, not in a really big way, but certainly in a big enough way to raise my curiosity as well as my maternal alarm. Among the biggest things to go wrong, she developed jaundice within a day of being born—but, then, so do about 50% of all full-term babies. Jaundice, characterized by a distinctive yellowish cast to the skin, is caused proximately by the inability of the liver to process the products of old, broken-down blood cells and excrete them properly. Hyperbilirubinemia refers to elevated levels of bilirubin, the principal breakdown product of hemoglobin, the oxygen-carrying pigment in red blood cells.

Usually, jaundice resolves itself without consequences but, in Hannah's case, she was just as jaundiced at four days as she was at one day of age. The family doctor prescribed phototherapy—exposing the baby to broad spectrum white light to photo-isomerize the bilirubin. Changing the structure of the bilirubin by exposing it to light energy often renders it more water-soluble and hence excretable. After a day of phototherapy, Hannah's bilirubin levels had dropped to the point that we were allowed to take her home. But those levels remained higher than normal and we watched helplessly as the home health care nurse came at regular intervals to stick needles into her tiny foot and collect blood for monitoring. About a month after she came home, we were told that her bilirubin levels were unacceptably high; at this point, we didn't need to see lab test results, because, two weeks past Halloween, our daughter was still looking about as orange as a pumpkin.

Despite the intriguing aesthetic possibilities presented by an orange baby, we were extremely worried; accumulation of bilirubin in the brain can cause developmental delay, hearing loss, learning disorders, and perceptual motor handicaps. So we were more than anxious to bring her in for additional testing. Watching her being poked and prodded was absolute agony, but, underlying the agony was an insatiable need to know what was going on inside my daughter. My entomological training was basically useless here. With a few truly bizarre exceptions (such as the chironomid bloodworms that breed in sewage), insects don't even have hemoglobin. They rely on a system of pipes and tubes for oxygen delivery, not an oxygen-carrying pigment. The tests were all uninformative, other than to indicate a problem with liver function. None of this was reassuring, and we were starting to get concerned about the long-term impact of the poking and prodding over and above the possible liver problem. One metabolic assay involved injecting Hannah with radioactive tracers and watching their progress through her system—leading my husband to speculate whether we might be able to change her diapers at night for the next few days without turning the lights on. The only definitive bit of evidence was the result of a test conducted at the Mayo Clinic on a sample sent by the pediatric gastroenterologist who had taken over Hannah's care. This test revealed that Hannah was heterozygous for a mutant allele of alpha-1-antitrypsin.

Alpha-1-antitrypsin is a protein found in the blood. Its main function is to inhibit enzymes that break down other proteins, particularly leucocyte elastase from neutrophils (not trypsin, as the name "anti-trypsin" might suggest). Neutrophil elastase is an enzyme that destroys bacteria and damaged cells in the lung; in the absence of adequate levels of AAT, elastase can proceed to destroy normal lung tissue as well. The normal protein is made up of a single chain of 394 amino acids, equipped with three large carbohydrate side chains. At least 90 variants of AAT have been described and the vast majority of these function normally. There are, however, a few mutations that affect function. People with these mutant variants experience alpha-1-antitrypsin deficiency— that is, they do not produce sufficient blood levels of this protein. The major clinical manifestations of inadequate supplies of AAT are liver disease and early-onset emphysema. AAT is manufactured in liver cells; the mutant forms of the protein cannot exit the cells and in some cases accumulate and destroy the cells, leading to cirrhosis and hepatitis. Emphysema and other lung complications reflect the susceptibility of the alveoli in the lungs to destruction by elastase, the enzyme disarmed by AAT.

AAT deficiency is an autosomal recessive disease that is actually quite common among Caucasians of European descent; it has been estimated that a deficiency gene is found in as many as one in every 2,500 people in the United States. In Hannah's case, she possessed one normal allele, designated M, and one mutant allele of the Z form. The mutant AATZ molecule differs from the normal molecule by only a single nucleotide base—so that the amino acid lysine is substituted for glutamate at position 342. Infants born who are homozygous for the Z allele—without a normal allele—often experience such profound liver damage that they need a transplant in early childhood. Heterozygotes—those with one normal allele—often live their whole lives without manifesting any symptoms.

My husband and I faced this news with mixed emotions. It was disturbing to know that Hannah faced an elevated risk of lung and liver disease over the course of her life but it was a relief to know that she was not profoundly ill nor was she likely to become so, at least from childhood liver disease. I guess our involvement with AAT could have ended there, except that, as a biologist, after hearing that AAT deficiency is an autosomal recessive gene, I knew Hannah had to have inherited her mutant allele from one of us. The fact that she was heterozygous was strongly suggestive that one of us was heterozygous and faced those same elevated risks. I don't know if it was a normal reaction, but I really wanted to know which one of us it was. So I asked if we could send our blood off to the Mayo Clinic and, surprisingly enough, in this era of managed health care, our physician arranged for it.

As you might have guessed, I'm the mutant parent—like Hannah, I'm heterozygous for the Z allele. It all made sense in retrospect. My paternal grandfather, Hyman Berenbaum, died at the age of 60—a classic case of early onset emphysema, exacerbated by smoking. Smoking, by the way, substantially elevates the risk of heterozygotes for acquiring emphysema; whereas a normal nonsmoker's probability of surviving to age 60 is 85%, the probability of an AAT-deficient nonsmoker surviving to that age is 60%, and the probability of an AAT-deficient smoker surviving to that age is an eye-popping 7%, according to the Center for Human and Molecular Genetics of the University of Medicine and Dentistry of New Jersey. My grandfather was a smoker for most of his (short) adult life. My grandmother, who took up smoking to spite my grandfather, died at age 65 of smoking-related heart disease, but that's another story. And, upon hearing the saga of Hannah's genetic constitution, my mother recalled that, as newborns, all three of us—my brother, my sister, and I—had jaundice. During the low-tech 1950s, though, the standard phototherapy for infants born with jaundice was not treatment under a special broad spectrum white light but sitting out in the sunshine. All three of us were born in spring or summer months, so sunshine was plentiful and obligingly photoisomerized our bilirubin when called upon to do so.

The fact that I had survived forty-odd years blissfully unaware of even possessing a genetic deficiency cheered us both greatly; it seemed that, of genetic disorders, this was not a really bad one to inflict on anyone. In fact, there may even be a positive side. Hannah now has a compelling genetic reason to stay away from cigarette smoking—over and above the usual risks of lung cancer and heart disease, she faces a greatly elevated chance of contracting a lethal form of emphysema. And one might argue that, with liver function somewhat less than optimal, she may be at increased risk of liver damage from drinking—a compelling reason for staying away from alcohol. A genetic disorder that keeps one's child away from cigarettes and alcohol isn't completely bad. Now, if I can just figure out a way to tell her that AAT deficiency means tattooes and body piercing are risky, too. . . .

Subpoenas envy

I like to read biographies, or at least certain kinds of biographies, because there's something very satisfying about sharing an experience or attitude with another human being. The kinds of biographies I most like to read are those about people with whom I have something, no matter how potentially inconsequential, in common. I like to read biographies of women, scientists, and Jews, for example. This predilection has led me to some unlikely literature. As I read it, for example, I couldn't help wondering who else might have been interested in the biography of Sylvan Nathan Goldman, the (Jewish) inventor of the shopping cart. I bought a copy of Kary Mullis' book, *Dancing Naked in the Mind Field*, because I'm interested in the lives of scientists. It seemed unlikely when I began the book that I'd find much common ground. Mullis is a world-renowned molecular biologist who revolutionized contemporary life with his development of the polymerase chain reaction, the biochemical trick that enables almost anyone with the appropriate laboratory to amplify tiny traces of DNA into quantities that can be sequenced and identified. I work on caterpillars that eat parsnip plants. Mullis won a Nobel Prize for his scientific contributions; I won the Distinguished Teaching

Award from the North Central Branch of the Entomological Society of America (and, when I received the plaque, my name was misspelled).

So, it was a delightful surprise to find that Dr. Mullis and I had shared a science-related experience—sort of. In Chapter 4, titled "Fear and Lawyers in Los Angeles," Mullis describes his experience as an expert witness in the murder trial of O.J. Simpson, considered by many to be the trial of the century (or at least the trial of the 1990s). He was asked by Barry Scheck and Peter Neufeld, very well-known law professors who were part of O.J. Simpson's defense team, to testify about the reliability of PCR-based DNA testing; several drops of blood found at the crime scene not belonging to either victim were crucial bits of evidence being used by the prosecution against Simpson. Mullis had had previous experience as an expert witness on this very subject and, indeed, as developer of the technique used to obtain such evidence, his credentials were certainly above reproach. In his 19-page account of his experience, Mullis describes his assessment of the evidence—basically, the evidence was tainted due to proce-dural errors in the collection, preparation, and analysis of both the crime scene sample and the specimen of Simpson's blood. Although he actually did attend the trial, and even had a few exchanges with O.J. himself, he was never asked to testify. Johnnie Cochran, the chief defense lawyer, thought the jury was "saturated on the DNA issue" after weeks of "tedious and technical testimony" and was already more or less convinced that the DNA evidence was ambiguous. Although Mullis reported that he was paid for his time, he declined to indicate just how much he was paid, claiming that "the amount is a professional secret" but adding

BUZZWORDS

that he "drove back to La Jolla in the same 1989 Acura Integra" he had driven to the trial.

So, what do O.J. Simpson and DNA evidence have to do with me? Nothing, really, except that, like Kary Mullis, I was asked to be an expert witness in a trial and, like Dr. Mullis, I never got to testify. In my case, though, I never even made it into the courtroom. My brush with the law, as it were, began in May of 1997, when I received a phone call from John Greenwood, assistant district attorney of Coles County, Illinois. The name sounded vaguely familiar, and John explained that he had been a student at University of Illinois seven years earlier and had taken my Entomology 105 general education course. He was calling to ask me what I knew about scabies, parasitic mites that infest human skin; somehow, scabies figured into a child custody dispute. I'm certainly no expert on scabies, but I figured he had called me because I was the only entomologist he knew (and here's another reason I'm no expert—scabies mites are not insects; they're actually arachnids, with eight legs as adults). I ran through everything I could think of off the top of my head and whatever I said must have satisfied him because he then asked if I would be willing to testify in court about scabies. Inasmuch as I had only once before seen the inside of a courtroom, when I served as a juror on a less than thrilling civil case involving a furnace installation (not even the trial of 1986, much less of the decade), I said I'd be happy to, as long as he recognized that I was not really an expert on scabies.

My lack of formal credentials didn't bother John; he seemed happy that I was willing to testify, and he seemed especially happy that I was willing to do so without compensation (I thought it might be an interesting experience and a nice excuse to visit Coles

County). Then he told me I'd be subpoenaed, and I panicked. Up to this point in time, I had always associated subpoenas with Mafiosi and other unsavory types; I was happy to come to court, of my own free will. I naively didn't think we needed to get the law involved in it. John assured me that it was just a routine procedure and that there was no cause for alarm. So I sat and waited for my first subpoena to arrive, frantically reading every article about scabies ever published in a scientific journal in the interim.

Actually, that subpoena never arrived. I made several calls to John Greenwood, increasing in frequency as the court date, December 5, approached. I hadn't heard from anyone, including the lawyer, Penny Dodson, who was supposedly handling the case. I knew all about scabies, but I sort of thought that maybe the lawyer would brief me as to the details of the case before I was called upon to testify. Finally, I received a call from Penny Dodson, around December 1; she told me that the case had been "continued," meaning that it had been rescheduled for another date—in April, 1998. So, I read some more about scabies and went back to waiting for a subpoena. One finally arrived, on February 26. I found a middle-aged gentleman wandering around the second floor of Morrill Hall, dressed in a suit coat and tie. Middle-aged gentlemen are not that unusual in Morrill Hall, but coats and ties are encountered only rarely; this gentlemen didn't look like an instrument salesman or a publishing company representative selling introductory biology texts (the people who are usually found in Morrill Hall wearing coats and ties), so I asked if I could help him find something. He replied that he was looking for May Berenbaum. When I told him that I was May Berenbaum, he handed me a sheet of paper, on the top of which in boldface capital letters was written "SUBPOENA."

Of course, everyone in my lab had seen Officer Bill Wascher wandering around the hall and, after he had departed, asked who he was. I explained that he was a deputy sheriff who was giving me a subpoena to appear in court as an expert witness. I could see the skepticism in their eyes. Although I knew full well that I had done nothing wrong, I felt guilty just having been served with a subpoena. I channeled that guilt into more literature review and waited for a phone call from the lawyer.

And no phone call came. On April 3, I received a letter from Cathy Porter, "Victim Witness Coordinator," stating that I would "not need to appear on April 16, 1998, as earlier subpoenaed." The case had apparently been continued once again. My suspicions were confirmed when Officer Wascher returned to Morrill Hall on June 12, confidently making his way directly to my office this time with another subpoena, requesting my appearance in Coles County Courthouse on July 16, 1998. We exchanged pleasantries and he went on his way, leaving me once again to explain to graduate students, postdoctoral associates, and colleagues why a sheriff's deputy had once again come to visit me in my office.

By this point, the novelty of being subpoenaed had diminished somewhat, and I was losing interest in mastering the scabies literature. I still hadn't spoken to the lawyer when I received a phone message about a week before the scheduled trial that, again, I wouldn't need to appear. On July 27, I got a phone call from the county sheriff's office; Officer Wascher wanted to know if I'd be willing to stop by on my way home and pick up yet another subpoena, for a court appearance in Coles County on August 24. I guess the novelty of my being subpoenaed had worn a bit thin for him as well. Never having had an occasion to visit there, I didn't even know where the sheriff's office was located. It turns out it was in the same building as the county jail, so, on a rainy summer afternoon I casually walked into the county jail as inconspicuously as I could, wondering if any of my graduate students, postdoctoral associates, and colleagues might inopportunely happen by.

I don't remember whether I received a phone call or a letter that August, informing me that the case had been settled. I still don't know what scabies had to do with it, and whether the people of the state of Illinois won or the defendant won. In retrospect, I suppose I really don't have much at all in common with Kary Mullis. He gained a lot more insight into the workings of our legal system from his experience than I did from mine. But, then, I probably learned a lot more about scabies. Ask me sometime—I'm dying to tell someone all about them.

References

Note: This bibliography is something of a hybrid and suffers from many of the problems biological hybrids can experience. On one hand, my book publishers assure me that books for the general public do not require extensive bibliographies—in fact, they can be viewed as liabilities, perhaps because they convey the impression to the reader that there will be homework assignments and book reports to follow. On the other hand, I feel a compulsive need to follow scientific format and dutifully report to the reader every excruciating detail about every source I consulted. As a compromise, this bibliography is not completely comprehensive, but it is a bit more detailed than the sort of "related readings" reference lists often found in popular books. So I expect no reader will be completely happy—if it's any consolation, neither am I nor is the publisher. The authors who are cited are probably happy, though. For readers in search of homework assignments, more comprehensive bibliographies can be found in the original versions of those essays that were previously published in *American Entomologist*.

BUZZWORDS

Alpha-Tocopherol, Beta-Carotene Cancer Prevention Study Group. 1994. The effect of vitamin E and beta carotene on the incidence of lung cancer and other cancers in male smokers. New Eng. J. Med. 330:1029-1035.

Anonymous. 1985. Milton Levine is anty-establishment: he's been king of the hill since inventing the ant farm 30 years ago. People Weekly 24:99 (October 28, 1985).

Anonymous. 1995. Their job's no picnic: ant hunters Afton and Kent Fawcett say it's only the bites that bug them. People Weekly 44: 57 (Sept. 11, 1995).

Appelo, T., and S. Williams. 1999. Get buffed up! TV Guide 47:10-11 (July 31, 1999).

Archibald, R.G., and H.H. King. 1919. A note on the occurrence of a coleopterous larva in the urinary tract of man in the Anglo-Egyptian Sudan. Bull. Ent. Res. 9:255-256.

Arens, W. 1979. The Man-Eating Myth: Anthropology and Anthropophagy. New York: Oxford University Press.

Bacon, K.B., B. A. Premack, P. Gardner, and T.J. Schall. 1995. Activation of dual T cell signaling pathways by the chemokine RANTES. Science 269:1727-1730.

Badia, L., and V. J. Lund. 1994. Vile bodies: an endoscopic approach to nasal myiasis. J. Laryngology Otology 108:1083-1985.

Baker, D. 1987. Foreign bodies of the ears and nose in children. Pediatr. Emerg. Care 3:67.

Barry, D. 1988. Dave Barry's Greatest Hits. New York: Fawcett Columbine.

Bartholomaeus Anglicus. 1231. *De Proprietatibus Rerum*. (Frankfurt 1701 reprint: Frankfurt a. M.: Minerva, 1964) (translated by John Friedman, Dept. English, UIUC).

Blum, M., L. Rivier, and T. Plowman. 1981. Fate of cocaine in the lymantriid *Eloria noyesi*, a predator of *Erthryoxylon coca*. Phytochemistry 11:2499-2500.

Bodenheimer, F. 1951. Insects as Human Food. The Hague: W. Junk.

Borror, D., and D. DeLong. 1954. Introduction to the Study of Insects. New York: Rinehart.

Brauman, A., M.D. Kane, M. Labat, and J.A. Breznak. 1992. Genesis of acetate and methane by gut bacteria of nutritionally diverse termites. Science 257:1384-1387.

Braxton Hicks, J. 1859. Further remarks on the organs of the antennae of insects. Transactions of the Linnean Society of London 22:383-400.

References

Brottman, M. 1998. Meat Is Murder! An Illustrated Guide to Cannibal Culture. New York: Creation Books.

Brown, S.M. 1986. Of mantises and myths. BioScience 36:421-423.

Carlet, G. 1888. De la marche d'un insecte rendu tetrapode par la suppression d'une paire de pattes. Comp. Rend. Acad. Sci. Paris 197: 565-566.

Carpenter, M.M. 1945. Bibliography of Biographies of Entomologists. Am. Midl. Natur. 33:1-116.

Causey, N.B., and D.L. Tiemann. 1969. A revision of the bioluminescent millipedes of the genus *Motyxia*. Proc. Am. Phil. Soc. 113:14-33.

Chase, M.W., D.E. Soltis, R.G. Olmstead, D. Morgan, D.H. Les, B.D. Mishler, M.R. Duvall, R.A. Price, H.G. Hills, Y.L. Qui, et al. 1993. Phylogenetics of seed plants: an analysis of nucleotide sequences from the plastid gene *rbcL*. Ann. Missouri Bot. Gard. 80:528-80.

Ciftcioglu, N., K. Altintas, and M. Haberal. 1997. A case of human orotracheal myiasis caused by *Wolhlfahrtia magnifica*. Parasitol. Res. 83: 34-36.

Cleveland, L.R. 1923. Symbiosis between termites and their intestinal protozoa. Proc. Nat. Acad. Sci. USA 9:424-428.

Cocatre-Zilgien, J.H., and F. Delcomyn. 1990. Fast axon activity and the motor pattern in cockroach legs during swimming. Physiol. Entomol. 15:385-392.

Cohen, M.B., M.A. Schuler, and M.R. Berenbaum. 1992. A host-inducible cytochrome P450 from a host-specific caterpillar: molecular cloning and evolution. Proc. Natl. Acad. Sci. USA 89:10920-10924.

Collins, N.M., and T.G. Wood. 1984. Termites and atmospheric gas production. Science 224:84-86.

Conchillam. 1828. The tests by which a real mermaid may be discovered. Mag. Nat. Hist. 1:106-108.

Cruden, D.L., and A.J. Markovetz. 1984. Microbial aspects of the cockroach hindgut. Arch. Microbiol. 138:131-139.

Crypton, Dr. 1984. Timid Virgins Make Dull Company. New York: Penguin Books.

Curtis, H. 1983. Biology. New York: Worth.

Dance, P. 1976. Animal Fakes and Frauds. Berkshire: Sampson Low.

Dangerfield, J.M., and D.K. Mosugelo. 1997. Termite foraging on toilet roll baits in semi-arid savanna, South-East Botswana (Isoptera: Termitidae). Sociobiology 30:133-143.

Darwin, C. 1859. (reprinted 1962). The Origin of Species. New York: Crowell-Collier.

BUZZWORDS

David, M., E. Petricoin III, C. Benjamin, R. Pine, M.J. Weber, and A.C. Larner. 1995. Requirement for MAP kinase (ERK2) activity in interferon alpha and interferon beta-stimulated gene expression through STAT proteins. Science 269:1721-1723.

Defoliart, G.R. 1989. The human use of insects as food and as animal feed. Bull. Ent. Soc. Amer. 35:22-36.

Dibble, C.E., and A.J.O. Anderson, trans. 1963. The Florentine Codex, a General History of the Things of New Spain, Book 11 (Earthly Things). Santa Fe, N.M.: School of American Research and the Museum of New Mexico.

Dictionary of Mnemonics (no author). 1972. London: Eyre Methuen.

Disney, R., and H. Kurahashi. 1978. A case of urogenital myiasis caused by a species of *Megaselia* (Diptera: Phoridae). J. Med. Entomol. 14:717.

Disney, R. 1985. The Japanese species of *Megaselia* (Dipt., Phoridae) responsible for urogenital myiasis is a new species. Entomol. Monthly Mag. 121: 261-263.

Dugich-Djordjevic, M.M., G. Tocco, D.A. Willoughby, I. Najm, G. Pasinetti, R. F. Thompson, M. Baudry, P. A. Lapchak, and F. Hefti. 1992. BDNF mRNA expression in the developing rat brain following kainic acid-induced seizure activity. Neuron 8:1127-1138.

Duncan, C.D., and G. Pickwell. 1939. The world of insects. New York: McGraw-Hill.

Dunkle, S.W. 1991. Head damage from mating attempts in dragonflies (Odonata: Anisoptera). Ent. News 102:37-41.

Ehlers, M.D., W.G. Tingley, and R.L. Huganir. 1995. Regulation of hippocampal transmitter release during development and long-term potentiation. Science 269:1730-1734.

Eisemann, C.H., W.K. Jorgensen, D.J. Merritt, M.J. Rice, B.W. Cribb, P.D. Webb, and M.P. Zalucki. 1984. Do insects feel pain?—A biological view. Experientia 40:164-167.

Emmet, A.M. 1991. The Scientific Names of British Lepidoptera: Their History and Meaning. Colchester (UK): Harley.

Epler, J.H. 1987. Revision of the Nearctic *Dicrotendipes* Kieffer 1913 (Diptera: Chironomidae). Evol. Monog. 9:1-102.

Eyre. 1972. Dictionary of Mnemonics. London: Methuen.

Fabre, J. H. 1991. The conjugal meal of the mantis. Harper's Magazine 283:32-33 (reprinted from E. W. Teale, The Insect World of J. Henri Fabre, Boston: Beacon Press).

References

Feinle, C., D. Grundy, and N.W. Read. 1995. Fat increases vection-induced nausea independent of changes in gastric emptying. Physiol. Behav. 58:1159-1165.

Ferreira, T. 1993. The amazing live sea monkeys. TCI 27:36-39.

Ferris, G. 1919. A remarkable case of longevity in insects (Hem., Hom.). Ent. News 30:27-28).

Feyereisen, R., J.F. Koener, D. Farnsworth, and D.W. Nebert. 1989. Isolation and sequence of cDNA encoding a cytochrome P-450 from an insecticide-resistant strain of the house fly, *Musca domestica*. Proc. Natl. Acad. Sci. USA 86: 1465-1469.

Fraenkel, G. 1932. Untersuchungen über die Koordination von Reflexen und automatisch nervosen Rhythmen bei Insekten. Zeitschrift für vergl. Physiol. 16:371-393.

Fraenkel, G. 1939. The function of the halteres of flies (Diptera). Proc. Zool. Soc. Lond. Ser. A. 109:69-78.

Fuiman, L.A., and R.S. Batty. 1997. What a drag it is getting cold—partitioning the physical and physiological effects of temperature on fish swimming. J. Exp. Biol. 200:1745-1755.

Fulton, B.B. 1941. A luminous fly-larva with spider traits (Diptera: Mycetophilidae). Ann. Entomol. Soc. Amer. 34:289-302.

Gauld, I.D. 1991. The Ichneumonidae of Costa Rica 1. Mem. Amer. Entomol. Institute 47:1-589.

Gilbert, P. 1977. Compendium of the Biographical Literature on Deceased Entomologists. Woodland Hills (CA): William Sabbot Natural History Books.

Gomi, H., T. Ikeda, T. Kunieda, S. Itohara, S. B. Prusiner, and K. Yamanouchi. 1994. Prion protein (Prp) is not involved in the pathogenesis of spongiform encephalopathy in zitter rats. Neuroscience Letters 166:171-174.

Gorham, J.R. 1974. Tests of mosquito repellents in Alaska. Mosquito News 34:409-415.

Gorton Linsely, E. 1943. Delayed emergence of *Buprestis aurulenta* from structural timbers. J. Econ. Ent. 36:348-349.

Gould, S.J. 1984. Only his wings remained. Natural History 93(9):10-18.

Gould, S.J. 1986. Glow big glowworm. Natural History 95(12):10-16.

Guinness Book of World Records. 1998. Stamford, CT: Guinness Media Inc.

Gunder, J.D. 1929. A state butterfly for California. Pan-Pacific Entomologist 6:88-90.

Guthörl, P. 1934. Die Arthropoden aus dem Karbon und Perm des Saar-Nahe-Pfalz Gebietes. Abh. Preuss. Geol. Landesanst. N.F. 164:1-219.

Gwynne, D., and D. Rentz. 1983. Beetles on the bottle: male buprestids mistake stubbies for females. J. Aust. Ent. Soc. 22:79-80.

Hackstein, J.H.P., and C. K. Stumm. 1994. Methane production in terrestrial arthropods. Proc. Nat. Acad. Sci. USA 91:5441-5445.

Hammer, P.E., W. Burd, D.S. Hill, J.M. Ligon, and K.H. van Pee. 1999. Conservation of the pyrrolnitrin biosynthetic gene cluster among six pyrrolnitrin-producing strains. FEMS Microbiology Letters 180:39-44.

Hammock, B.D., B.F. McCutchen, J. Beetham, P.V. Chaudhary, E. Fowler, R. Ichinose, V.K. Ward, J.M. Vickers, B.C. Bonning, L.G. Harshman, D. Grant, T. Uematsu, and S. Maeda. 1993. Development of recombinant viral insecticides by expression of an insect-specific toxin and insect-specific enzyme in nuclear polyhedrosis viruses. Arch. Insect Biochem. Physiol. 22:315-344.

Harris, T. 1988. The Silence of the Lambs. New York: St. Martin's Press.

Harrison, R.J., and W. Montagna. 1969. Man. New York: Appleton-Century Crofts.

Haupt, H. 1949. Rekonstruktion permokarbonischer Insekten. Beitr. Taxon. Zool. 1:23-42.

Heard, H.F. 1941. A Taste for Honey. New York: Vanguard Press.

Heinonen, O.P., et al. 1994. The effect of vitamin E and beta carotene on the incidence of lung cancer and other cancer in male smokers. New England Journal of Medicine 330:956-961.

Herrick, G.W. 1926. The "ponderable" substance of aphids (Homop.). Entomol. News 37:207-211.

Herzog, A. 1974. The Swarm. New York: New American Library.

Hinton, H.E. 1951. A new Chironomid from Africa, the larva of which can be dehydrated without injury. Proc. Zool. Soc. Lond. 121:371-380.

Hinton, H.E. 1960a. Cryptobiosis in the larva of *Polypedilum vanderplanki* Hint. (Chironomidae). J. Insect Physiol. 5:286-300.

Hinton, H.E. 1960b. A fly larva that tolerates dehydration and temperatures of −270° to +102°. Nature 188: 337-338.

Hodge, C.F., and J. Dawson. 1918. Civic Biology. Boston: Ginn.

Holt, V.M. 1885. Why not eat insects? Repr. Hampton: E. Classey.

Howard. L.O. 1886. The excessive voracity of the female Mantis. Science 8: 326.

Howard. L.O. 1894. A review of the work of the Entomological Society of Washington during the first ten years of its existence. Proc. Ent. Soc. Wash. 3:161-167.

References

Howard. L.O. 1911. The House Fly—Disease Carrier. New York: Stokes.

Howard. L.O. 1931. The Insect Menace. New York: Century.

Howard. L.O. 1934. More about the beginnings of the Society. Proc. Ent. Soc. Wash. 36:51-55.

Howell, M., and P. Ford. 1985. The Beetle of Aphrodite and Other Medical Mysteries. New York: Random House.

Hu, S., R. Stritzel, A. Chandler, and R.M. Stern. 1995. P6 acupressure reduces symptoms of vection-induced motion sickness. Aviation, Space, and Environmental Medicine 66:631-634.

Huang, Z.J., I. Edery, and M. Rosbash. 1993. PAS is a dimerization domain common to *Drosophila* period and several transcription factors. Nature 364:259-262.

Hudson, G.V. 1886. A luminous insect larva in New Zealand. Entomol. Monthly Magazine 22:99-100.

Hurd, P. D. 1954. Myiasis resulting from the use of the aspirator method in the collection of insects. Science 119:814-815.

Hurter, J., E.F. Boller, E. Stadler, B. Blattman, H.R. Buser, N.V. Bosshard, L. Damm, M.W. Kozlowksi, R. Schoni, et al. 1987. Oviposition deterring pheromone in *Rhagoletis cerasi* L.: Purification and determination of chemical constitution. *Experientia* 43:157-164.

Hutchison, D.C.S. 1998. α-1-Antitrypsin deficiency in Europe: geographical distribution of P1 types S and Z. Resp. Med. 82:367-377.

International Code of Zoological Nomenclature. 1985. Berkeley, Calif.: University of California Press.

Johnstone, I., and J. Cleese. 1997. Fierce Creatures. New York: Boulevard Books.

Jordan, D.S., and V.L. Kellogg. 1908. Evolution and animal life. New York: Appleton.

Kahle, J.B. 1985. A view and a vision: women in science today and tomorrow. In J.B. Kahle, ed. Women in Science: A Report from the Field. Philadelphia: Falmer Press, pp. 193-195.

Kandel, E.R., and J. H. Schwartz. 1985. Principles of Neural Science. Second Edition. New York: Elsevier.

Kearfoot, W.D. 1907. New North American Tortricidae. Trans. Amer. Ent. Soc. 33:1-94.

Khalil, M.A.K., and R.A. Rasmussen. 1983. Termites and methane. Nature 302:355.

Kirkaldy, G.W. 1904. An addition to the rhynchotal fauna of New Zealand. Entomologist 37: 279-283.

BUZZWORDS

Knols, B.G.J., J.A.J. van Loon, A. Cork, R.D. Robinson, W. Adam, J. Meierjink, R. DeJong, and W. Takken. 1997. Behavioural and electrophysiological responses of the female malaria mosquito *Anopheles gambiae* (Diptera: Culicidae) to Limburger cheese volatiles. Bull. Ent. Res. 87:151-159.

Kutz, F.W. 1974. Evaluations of an electronic mosquito repelling device. Mosquito News 34:369-375.

Kynaston, S.E., P. McErlain-Ward, and P.J. Mill. 1994. Courtship, mating behaviour, and sexual cannibalism in the praying mantis, *Sphodromantis lineola*. Anim. Behav. 47:739-741.

Langmuir, I. 1938. The speed of the deer fly. Science 87:233-234.

Larson, G. 1987. Hound of the Far Side. New York: Andrews, McMeel and Parker, p. 56.

Larson, G. 1989. A Prehistory of the Far Side: a 10th Anniversary Exhibit. Kansas City: Andrews and McMeel.

Lawrence, S. E. 1992. Sexual cannibalism in the praying mantid, *Mantis religiosa*: a field study. Anim. Behav. 43:569-583.

Leffler, S., P. Cheney, and D. Tandberg. 1993. Chemical immobilization and killing of intra-aural roaches: an in vitro comparative study. Ann. Emerg. Med. 14:1795-1798.

Ligett, H. 1931. Parasitic infestations of the nose. J. Amer. Med. Assoc. 96:1571-1572.

Lillehoj, E.B., W.F. Kwozek, M.S. Zuber, A.J. Bockhott, O.H. Calvert, W.R. Findley, W.D. Guthrie, E.S. Horner, and L.M. Josephson. 1980. Aflatoxin in corn (*Zea mays*) before harvest: Interaction of hybrids and locations. Crop Science 20:731-734.

Lillehoj, E.B., D.J. Fennell, W.F. Kwolek, G.L. Adams, M.S. Zuber, E.S. Horner, N.W. Widstrom, H. Warren, W.D. Guthrie, D.B. Salier, W.R. Findley, A. Manwiller, L. M. Josephson, A. J. Bockholt. 1978. Aflatoxin contamination of corn before harvest: *Aspergillus flavus* association with insects collected from developing ears: *Ostrinia nubilalis, Heliothis zea, Spodoptera frugiperda*. Crop Sci. 18:921-924.

Lillehoj, E.B., W.F. Kwolek, M.S. Zuber, E.S. Horner, N.W. Widstrom, W.D. Guthrie, M. Turner, D.B. Salier, W.D. Findley, A. Manwiller, and L.M. Josephson. 1980. Aflatoxin contamination caused by natural fungal infection of preharvest corn. Plant and Soil 54:469-475.

Lipton, B. 1991. Bug Busters. Garden City Park, N.Y.: Avery Publishing Group, Inc.

References

Liske, E., and W. J. Davis. 1984. Sexual behaviour of the Chinese praying mantis. Anim. Behav. 32:916-944.

M.C.G. 1829. Queries and answers—a sea spider. Mag. Nat. Hist. 2:211.

Marcus, N., J.H. Teckman, and D.H. Perlmutter. 1998. α-1-Antitrypsin deficiency: from genotype to childhood. J. Ped. Gastroent. Nutri. 27:65-74.

Martin, L. 1997. The X Files #10: Die, Bug, Die! New York: Harper Collins.

Martius, C., R. Wassman, U. Thein, A. Bandeira, H. Rennenberg, W. Junk, and W. Seiler. 1993. Methane emission from wood-feeding termites in Amazonia. Chemosphere 26(1-4): 623-632.

Matsuki, N., C.-H. Wang, F. Okada, M. Tamura, Y. Ikegaya, S.-C. Lin, Y.-N. Hsu, L.-J. Chaung, S.-J. Chen, and H. Saito. 1997. Male/female differences in drug-induced emesis and motion sickness in *Suncus murinus*. Pharmacol. Biochem. Behav. 57:721-725.

McGehee, D.S., M.J.S. Heath, S. Gelber, P. Devay, and L.W. Role. 1995. Nicotine enhancement of fast excitatory synaptic transmission in CNS by presynaptic receptors. Science 269:1692-1696.

Melander, A.L. 1924. Review of the dipterous family Piophilidae. Psyche 31:78-86.

Menke, A. 1977. *Aha*, a new genus of Australian Sphecidae, and a revised key to the world genera of the tribe Miscophini (Hymenoptera, Larrinae). Bull. Ent. Soc. Poland 47:671

Menke, A. 1993. Biological and Other Generally Unsupported Statements. B.O.G.U.S. 2:24-27.

Merck Manual of Diagnosis and Therapy, R. Berkow, ed. 1982. Rahway: Merck Sharp & Dohme Research laboratories.

Messersmith, D.H. 1976. Long live the entomologist! Insect World Digest 3:21.

Metcalf, C.L., and W.P. Flint. 1928. Destructive and Useful Insects. New York: McGraw Hill.

Meyrick, E. 1886. A luminous insect larva in New Zealand. Entomol. Monthly Mag. 22:266-267.

Millar, T. P. 1986. Humor: the triumph of reason. Persp. Biol. Med. 29:545-558.

Morris, D. 1991. A taste for their own kind. National Wildlife 29:14-16.

Mullis, K. 1998. Dancing Naked in the Mind Field. New York: Pantheon Books.

Murray, K.D., A.R. Alford, E. Groden, F.A. Drummond, R.H. Storch, M.D. Bentley, and P.M. Sugathapala. 1993. Interactive effects of an antifeedant used with *Bacillus thuringiensis* var. San Diego delta endotoxin on Colorado potato beetle (Coleoptera: Chrysomelidae). J. Econ. Ent. 86:1793-1801.

BUZZWORDS

Nalcaci, O.B., and A.N. Bozcuk. 1990. The effect of vitamin E treatment on the life-span of *Drosophila*. Doga Biyoloji Serisi 14(3):157-163.

Neal, J., and M. Berenbaum. 1989. Decreased sensitivity of mixed-function oxidases from *Papilio polyxenes* to inhibitors from host plants. J. Chem. Ecol. 15:439-446.

Nelkin, D. 1992. Living inventions: biotechnology and the public. Nat. Ag. Biotech. Council 4:63-72.

Nelson, D., et al. 1993. The P450 superfamily: update on new sequences, gene mapping, accession numbers, early trivial names of enzymes, and nomenclature. DNA Cell Biol. 12:1-51.

Neumoegen, B. 1893. Description of a peculiar new liparid genus from Maine. Can. Ent. 25:213-215.

Nolch, G. 1997. Perfectly inflated genitalia every time. Search 28:107.

Norman, M. R., A.P. Mowat, and D.C.S. Hutchison. 1997. Molecular basis, clinical consequences and diagnosis of alpha-1-antitrypsin deficiency. Ann. Clin. Biochem. 34:230-246.

Normile, D. 1996. Global interest high, knowledge low. Science 274:1074.

Office of Technology Assessment, 1988. Educating Scientists and Engineers: Grade School to Grad School. Washington, DC.: Congress.

Oldroyd, H. 1964. The Natural History of Flies. New York: Norton.

Osten-Sacken, C.R. 1886a. A luminous insect-larva in New Zealand. Entomol. Monthly Mag. 23:133-134.

Osten-Sacken, C.R. 1886b. More about the luminous New Zealand larvae. Entomol. Monthly Mag. 23:230-231.

O'Toole, K., P.M. Paris, and R.D. Stewart. 1985. Removing cockroaches from the auditory canal: a controlled trial. New England J. Med. 312:1192.

Partsch, C.J., R. Huemmelink, M. Peter, W.G. Sippel, W. Oostdijk, R.J.H. Odink, S.L.S. Drop, et al. 1993. Comparison of complete and incomplete suppression of pituitary gonadal activity in girls with central precocious puberty: Influence on growth and predicted final height. Horm. Res. 39:111-117.

Partridge, E. 1974. A Dictionary of Slang and Unconventional English. New York: Macmillan Pub. Co., Inc.

Paulson G., and R.D. Akre. 1994. A fly in ant's clothing. Nat. Hist. 1/94:56-58.

Perlmutter, D.H. 1998. Alpha-1-antitrypsin deficiency. Seminars in Liver Disease 18:217-225.

Phelan, M.J., and M.W. Johnson. 1995. Acute posterior ophthalmomyiasis treated with photocoagulation. Am. J. Ophthamol. 119:106-109.

References

Pitzalis, A., G. Borgioli, and G. Messana. 1991. The organ of Bellonci in the family Stenasellidae (Isopoda: Asellota): Light microscope results for the genus *Stenasellus*. Memoires de Biospéleologie 18:155-158.

Porterfield, A. L. 1987. Does sense of humor moderate the impact of life stress of psychological and physiological well-being? J. Res. Personality 21:306-317.

Powell, J. 1989. Synchronized mass-emergences of a yucca moth, *Prodoxus y-inversus* (Lepidoptera: Prodoxidae) after 16 and 17 years in diapause. Oecologia 81:490-493.

Rasmussen, R.A., and M.A.K. Khalil. 1983. Global production of methane by termites. Nature 301:700-702.

Rau, P. 1945. Food preferences of the cockroach, *Blatta orientalis*. Ent. News 56:276-278.

Rau, P. 1945. Longevity as a factor in psychic evolution. Annals of the Entomological Society of America 38:503-504.

Raubenheimer, D., and S. J. Simpson. 1992. Analysis of covariance: an alternative to nutritional indices. Entomologia Experimentalis et Applicata 62:221-231.

Raubenheimer, D., and S.J. Simpson. 1994. The analysis of nutrient budgets. Functional Ecology 8:783-791.

Riley, C. V., and L.O. Howard. 1893. The female rear-horse versus the male. Insect Life 5:145.

Roeder, K.D. 1935. An experimental analysis of the sexual behavior of the praying mantis *Mantis religiosa* L. Biol. Bull. 49:203-220.

Roeder, K.D. 1967. Nerve Cells and Insect Behavior. Cambridge: Harvard University Press.

Schittek, A. 1980. Insects in the external auditory canal—a new way out. JAMA 243:331.

Schmidt, M.H., K. Sakai, J-L. Valatx, and M. Jouvet. 1999. The effects of spinal or mesencephalic transections on sleep-related erections and ex-copula penile reflexes in the rat. Sleep 22:409-418.

Seki, K., and M.Toyoshima. 1998. Preserving tardigrades under pressure. Nature 395:853-854.

Seymour, W., and N. M. Hewitt. 1997. Talking about Leaving: Why Students Leave the Sciences. Boulder: Westview Press.

Shammi, P., and D. T. Stuss. 1999. Humour appreciation: a role of the right frontal lobe. Brain 122:657-666.

Sharma, H., D. Dayal, and S.P. Agrawal. 1989. Nasal myiasis: review of 10 years experience. J. Laryngology Otol. 103:489-491.

BUZZWORDS

Simmons, P. 1927. The Cheese Skipper as a pest in cured meats. Bull. U.S. Dept. Agric. 1453:1-55.

Sivinski, J.M. 1981. "Love bites" in a lycid beetle. Fla. Ent. 64:541.

Sivinski, J.M. 1998. Phototropism, bioluminescence, and the Diptera. Florida Entomologist 81:292-392.

Snodgrass, R. E. 1933. Insect Morphology. New York: McGraw-Hill Book Co.

Spilman, T.J. 1984. Vignettes of 100 years of the Entomological Society of Washington. Proc. Ent. Soc. Wash. 86:1-10.

Spratt, E.C. 1980. Male homosexual behavior and other factors influencing adult longevity in *Tribolium castaneum* and *Tribolium confusum*. J. Stored Prod. Res. 16:109-114.

Steele, L.P., E.J. Dlugokencky, P.M. Lang, P.P. Tans, R.S. Martin, and K.A. Masarie. 1992. Slowing down of the global accumulation of atmospheric methane during the 1980s. Nature 358:313-316.

Szuromi, P. 1995. This week in *Science*. Science 269:1649.

Thackrah, C.T. 1832. The effects of arts, trades, and professions, and of civic states and habits of living, on health and longevity. London: Longman, Rees, Orme, Brown, Green and Longman. (reprinted by Canton, Mass: Watson Publishing, 1985).

Tomita, M., Y. Uchijima, K. Okada, and N. Yamaguchi. 1984. A report of self-amputation of the penis with subsequent complication of myiasis. Hinyokika Kiyo 30:1293-1296.

Townsend, C. 1927. On the Cephenemyia mechanism and the daylight-day circuit of the earth by flight. J. New York Ent. Soc. 35:245-252.

Transcience Corporation. 1987. It's Fun to Raise Pet Sea Monkeys: Official Sea-Monkey Handbook. Bryans Road, Md.: Transcience Corporation.

Treisman, M. 1977. Motion sickness: an evolutionary hypothesis. Science 197:493-495.

Tsugorka, A., E. Rios, and L.A. Blatter. 1995. Imaging elementary events of calcium release in skeletal muscle cells. Science 269:1723-1726.

V. 1829. Notice of an imposture entitled a pygmy bison, or American ox. Mag. Nat. Hist. 2:218-219.

von Bloeker, Jr., J.C. 1976. California was first! Insect World Digest 3:17.

von Buddenbrock, W. 1921. Der Rhythmus der Schreitbewegungen der Stabheuschrecke *Dixippus*. Biol. Zentralbl. 41:41-48.

References

Wadhwa, R., M. Kaur, and S.P. Sharma. 1988. An antioxidant induced alteration in peroxidae activity in aging *Zaprionus paravittiger* (Diptera). Mech. Aging Dev. 45:277-283.

Waldbauer, G.P. 1968. The consumption and utilization of food by insects. Advances in Insect Physiology 5:229-288.

Wang, J.-J., and M.B. Dutia. 1995. Effects of histamine and betahistine on rat medial vestibular nucleus neurones: possible mechanism of action of anti-histaminergic drugs in vertigo and motion sickness. Exp. Brain Res. 105: 18-24.

Warren, J., and L.C. Rotello. 1989. Removing cockroaches from the auditory canal: a direct method. New Eng. J. Med. 320:322.

Weiss, H.B. 1945. How long do entomologists live? Ent. News 56:189-190.

Wells, H.G. 1974. The Complete Short Stories of H. G. Wells. New York: St. Martin's Press, Inc.

Welsch, R. 1976. Tall-Tale Postcards. New York: A.S. Barnes.

Wheeler, W.M., and F. X. Williams. 1915. The luminous organ of the New Zealand glow-worm. Psyche 22:36-43.

Wigglesworth, V.B. 1980. Do insects feel pain? Antenna 4:8-9.

Wilcke, J.T.R. 1998. Late onset genetic disease: where ignorance is bliss, is it folly to inform relatives? Brit. Med. J. 317:744-747.

Wilson, E. O. 1995. Naturalist. Washington: Island Press.

Yager, J. 1989. *Pleomothra apletocheles* and *Godzilliognomus frondosus*, two new genera and species of remipede crustaceans (Godzilliidae) from anchialine caves of the Bahamas. Bulletin of Marine Science 44(3):1195-1206.

Yuswasdi, C. 1950. Tinnitus aurium and the luminous milliped. Siriraj Hospital Gazette 2:194.

Zimmerman, P.R., J.P. Greenberg, S.O. Wandiga, and P.J. Crutzen. 1982. Termites: a potentially large source of atmospheric methane, carbon dioxide, and molecular hydrogen. Science 218:563-565.

Zimmerman, P.R., and J.P. Greenberg. 1983. Termites and methane. Nature 302:354-356.

Zimmerman, P.R., J. P. Greenberg, and J.P.E.C. Darlington. 1984. Termites and atmospheric gas production. Science 224:86.

Zimmerman, P.R., J. P. Greenberg, and J.P.E.C. Darlington. 1996. World Almanac and Book of Facts. New York: World Almanac Books.

BUZZWORDS

Web sources

Georgia Tech alumni Web page (www.alumni.gatech.edu/news/ttopics/sum98/
yellowjackets.html)

www.eudesign.co.uk/mnems/_mnems.htm

NSF Gallup poll: www.nsf.gov/od/lpa/nstw/teenov.otm

www.walshnet.com/walshnet/punster/puns.html

http://pc159.lang.uiuc.edu/class.pages/DAAD96/TV/ourpage

Index

A

Aaages sp., 158
Acronyms, 212-217
Adrenalin O.D., 80
Advertising/commercials, insects in, 119-120, 131
Aeshnid dragonflies, 17
Agra vation, 160
Agriculture, and termites, 10
Aha ha, 159
Airspeed of insects, 40-42
Akens, Jewel, 77
Aldrin, Buzz, 191
Aldrovandi, Ulysse, 181
Allen, Fred, 250
Alpert, Herb, 77
Alpha-1-antitrypsin deficiency, 257-259
Amber, insects in, 14
American cockroach, 12
American Entomological Institute, 164
American Entomologist, 183, 219
American Journal of the Crush-Freak, 18-20

American Museum of Natural History, 224, 225, 228
Amphipods, 51
Amyotrophic lateral sclerosis, 209
Andries, 169-170
Angelicus, Bartholomaeus, 168-169
Animal Fakes and Frauds, 174-175
Annals of the Entomological Society of America, 185
Annals of the Missouri Botanical Gardens, 209
Anopheles gambiae, 222-223
Ant, 18, 31, 238
    burning with magnifying glass, 147
    eponymous body parts, 154
    farms, 45-46, 147
    festival, 128
    games related to, 84, 85
    longevity, 3-5, 6, 7
    Microdon cohabitation with, 169, 172
    movies, 201
    psychotropic effects, 123-124
    in songs, 113
    superheroes, 66, 68, 70

BUZZWORDS

Index

BUZZWORDS

Index

BUZZWORDS

Index

Index

Index

Index

BUZZWORDS

Index

BUZZWORDS

Index

BUZZWORDS